Discovering the Universe

The Story of Astronomy

Paul Murdin

ANDRE DEUTSCH

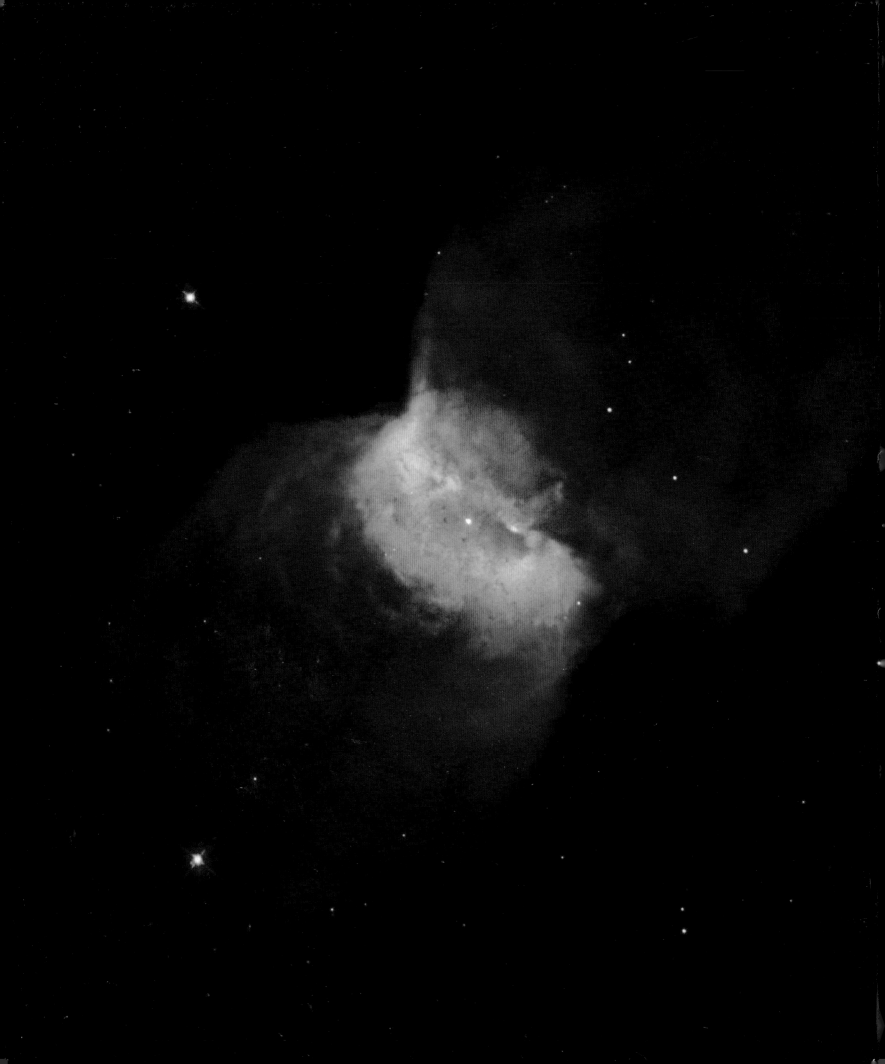

Contents

Introduction 5
1   Monuments and calculators 6
2   Eyes on the sky 12
3   Patterns in the sky – the beginning of astronomy 14
4   The nature of the Earth-centred universe 20
5   Astronomy in the dark 24
6   Revolutions of the heavenly spheres: Copernicus 28
7   Measuring the skies from the City of the Stars 30
8   Galileo Galilei 34
9   Cosmographic mysteries 42
10 Universal attraction 48
11 New eyes on the sky 54
12 New planets 56
13 The stars 66
14 The life of stars 72
15 The death of stars 78
16 New windows on the universe 82
17 Exploding stars 86
18 The origin of the chemical elements 90
19 The birth of stars and planets 94
20 Exploration of the planets 98
21 The inner planets 102
22 The Earth, the Moon and Mars 108
23 The gas giants 116
24 Other planetary systems 124
25 Our Galaxy and others 126
26 The universe of galaxies 132
27 Exploding galaxies and quasars 140
28 The expanding universe 144
29 Dark matter and dark energy 148
30 Life in the Universe 152
Index 156
Credits 160

THIS IS AN ANDRE DEUTSCH BOOK

This edition published in 2014 by Andre Deutsch Limited
a division of the Carlton Publishing Group
20 Mortimer Street
London W1T 3JW

Text © Paul Murdin, 2011, 2014
Design © Carlton Books, 2011, 2014

All rights reserved. This book is sold subject to the condition that it may not be reproduced, stored in a retrieval system or transmitted in any form or by any means, electronic, mechanical, photocopying, recording or otherwise without the publisher's prior consent.

A CIP catalogue record for this book is available from the British Library.

First published in 2011 as *Mapping the Universe*

ISBN 978 0 233 00442 6

Introduction

The night sky is a common heritage of mankind. Whoever and wherever we are, the sun, the stars and the planets shine down on us all, provoking wonder and inspiring questions about our place in the scheme of things. In contemplating the stars, we focus outwards and we are drawn out of our mundane predicament, at least for a time. Setting aside our differences with our fellow human beings, we can stand, gazing up, shoulder to shoulder, not eyeball to eyeball.

So it has been for thousands of years. There is evidence that our ancestors as far back as the Ice Ages were interested in astronomy, noting the patterns made by the constellations and the rhythms of the seasons, the moon and the planets. Our ancestors learnt how astronomy provided essential knowledge that helped keep time and the calendar. More subtly, astronomy formed the backdrop to our ancestors' world picture, at first a cosy, almost infantile view that the universe is constant and reliable, organized around our human needs; later came the modern realization that we humans are a minute, transient part of a much larger whole, packed with eventual or sporadic disasters. The inspiring human reaction to this picture, which could be a terrifying and paralysing one, is that we have come to accept this reality with equanimity, and to investigate its ramifications with even greater intensity. Thus we are living now in a golden age of astronomy, having passed beyond the technical investigation of the arcane mathematics of orbits to the hopeful contemplation of some of the major questions of existence. How did it all start? How will it all end? Are we alone in the universe? We can tackle some of these questions scientifically. We have started to deliver at least some aspects of the answers.

This book is about the growing development of our understanding of the universe through the centuries. It is about the way that technological advances stimulated astronomy and astronomy stimulated technological development, the one leapfrogging the other in a constant progression. But, of course, science is only in part about paraphernalia; it is mostly about people: their false starts, their mistakes, their sudden inspirations, their logical deductions and their vision. In this book I tell the history of astronomy through the astronomers, the way they thought and the growth of their understanding of the universe in which we live. I have been proud to have been an astronomer throughout my working life, and I feel an affinity with my predecessors. I hope to have told their stories with respect and affection.

1

Monuments and calculators

From time immemorial, people have looked up with wonder and awe to the starry sky above. As long ago as the Stone Age, people formed the random positions of stars into patterns in the sky – the constellations – and projected on to them visions of their gods. They also noticed the regular changes in the positions of some of the brightest heavenly bodies – the planets.

Physical evidence that prehistoric people had an interest in astronomy survives in some of their buildings. At the very dawn of history, they built expensive monuments to study and to measure the stars and planets and, fearing that even the most powerful among them were subject to their rule, they sought to predict where the planets would be in the future. The orientation of these buildings relative to the Sun might be understood as simply the use of natural light and warmth, but the structures were also used by a priestly class of astrologers to make calendrical calculations needed for practical and ritual ceremonies.

Stonehenge is perhaps the most famous prehistoric astronomical structure. It is a megalithic monument in southern England. Its principal feature – the "henge" that gives it its name – is a circle of megaliths (large stones) on the tops of which are balanced other stones looking like lintels in a succession of doorways. Other structures – stones, ditches, wooden palisades – formed concentric circles around the centre of the henge. Stonehenge is the centre-piece of a landscape of prehistoric barrows (or tombs), and was approached by a trackway that runs from east to west. The stone circle has openings and alignments that are oriented toward sunrise in midsummer and midwinter.

Stonehenge was built between 3100 and 1600 BC. This long usage, the labour that was needed, the landscape in which it lies, and the fact that it was never lived in imply that Stonehenge was used to commemorate important dead people at ritually important seasons chosen from the solar calendar.

In 1963 the astronomer Gerald Hawkins (1928–2003) argued that numerous alignments in Stonehenge could have been used to predict eclipses. Hawkins's theories were developed by the cosmologist Fred Hoyle (1915–2001). These elaborate theories have been criticized by archaeologists, but there is little doubt about the main alignment, which was first noted in 1740 by the antiquary William Stukeley (1687–1765).

Right *Stonehenge, in the south of England, is an elaborate system of prehistoric circles of massive stones that are oriented towards sunrise.*

Similar alignments are found in other megalithic monuments in the British Isles and Brittany in France. One example is the passage tomb at Newgrange, County Meath, Ireland, in which there is a "lightbox", an aperture which floods the interior of the tomb with sunlight at the moment of midwinter sunrise (winter solstice).

Sun, stars and stones

The Pyramids at Giza in Egypt also lie on a precise alignment, being square on to the cardinal points. This orientation must have been established astronomically. The temple of the sun-god Amun-Re at Karnak near Thebes was oriented towards the rising of the midwinter sun at the time of its construction; indeed the entire complex is aligned with respect to the Sun.

The people of the Chaco Canyon culture of Native American people in New Mexico, which flourished between AD 900–1150, observed the rising of the Sun behind the mountains on the horizon, as seen from special observation platforms. In this way, they kept track of the seasons, enabling them to time agricultural activities and ritual ceremonies. The same calendrical skill survives even now in the prehistoric people's descendants, the Hopi Indians. There are numerous similar prehistoric sites in the Southwest of the United States and Northwestern Mexico, where there are solar alignments that mark the calendar. It is understandable that an awareness of the calendar would have been developed in an area where the growing season is short, and it is critical to plant crops like maize at the right moment so they have time to ripen.

In the same general area in 1950, astronomers Helmut Abt and Bill Miller found pictographs in White Mesa and Navajo Canyon that represented a completely different phenomenon. The drawings or paintings of the crescent Moon and a bright star, made by Pueblo people – ancestral Native Americans – seem to show the supernova of 1054. It is known from Chinese records that this appeared on 4 July 1054 with the crescent Moon nearby. A similar pictograph at Pueblo Bonito in Chaco Canyon, New Mexico, shows the star and crescent and a hand-print, and may also represent the supernova.

In South America, according to early Spanish colonists,

Left *In Chaco Canyon, New Mexico, a pictograph depicts a star, interpreted as the supernova of 4 July 1054, as it was in the dawn sky that morning with the crescent moon. The painter "signed" the pictograph with a handprint.*

Pages 10–11 *Among the designs in the Nazca Desert of Peru, a spider is controversially thought by Dr Phillis Pitluga of the Adler Planetarium to be a diagram of the constellation that we call Orion.*

THE ASTRONOMY OF THE MAYAS

The Maya people of Mexico had very detailed knowledge of the movements of the planets. Uxmal is a Mayan city in the Yucatán Peninsula, built about AD 500–1100. The so-called Governor's Palace is aligned with an azimuth of 118°, the extreme southerly rising of the planet Venus which occurs once every eight years. In isolation, this coincidence would not seem significant, were it not for the fact that the façade of the palace is covered with stone carvings that represent Venus and zodiacal constellations. It is also well known to ethnologists that the Mayan civilization used a calendar based on the motions of Venus. The oldest surviving book from the Americas (from around the eleventh century AD, but based on material dating to several hundred years earlier) is the *Dresden Codex*, which is an astronomical compendium describing how to make calendrical calculations using seasonal, medical and religious and other astrological information. Another Mayan monument, El Castillo, or Kukulcán's Pyramid, is a step-pyramid in Chichen Itza and has 365 steps, one for each day of the year, and, like other buildings in the complex, astronomical alignments oriented to the Sun, Moon and Venus.

Below Standing on a stepped tier on the upper floor of the Governor's Palace in the Mayan ruins at Uxmal in Mexico, the façade of the uppermost building is covered with designs of astronomical significance.

Above According to astronomer Norman Lockyer (1836–1920), the Great Hall in the temple of Amun-Re at Karnak was built by Ramses II on an alignment towards the setting of the midsummer Sun. The picture is the hypostyle of the hall lithographed by Egyptologist Richard Lepsius in 1843.

the Incas of Peru used astronomical observations as a calendar. At Cusco, the capital city of the Inca empire, the Incas constructed ceques, radial routes emanating from the Coricancha (Temple of the Sun) at the centre of the city towards markers on the horizon called huacas, both natural and constructed for the purpose. There are alignments in all directions, whose significance could have been overlooked if it were not for the written ethnographic records.

The Nazca Lines are drawings in the Nazca Desert of Peru, dating from AD 400–650, and formed by brushing aside the stones lying on the desert surface. The drawings represent birds, mammals, spiders and other animals, and individual straight lines. According to a controversial theory by archaeologist Maria Reiche (1903–98), the lines were constructed as a solar calendar.

Eyes on the sky

Before the invention of the telescope, astronomical instruments to measure the positions of the stars and the planets were, fundamentally, simple sighted sticks. The simplest of all was the cross-staff, used to measure the angle above the horizon of a star (such as Polaris) or other body (perhaps the Sun), for navigational purposes. The navigator held one end of the stick against his cheek and sighted it on the horizon. He slid a cross-piece (or transom) along the stick until its top was in line with the star and read the angle between the star and the horizon from the position of the cross-piece on the stick.

Right *A surveyor uses a cross-staff (or Jacob's staff) in this German woodcut of 1531 from a manual on the instrument by Jacob Köbel.*

Below *A huge armillary sphere of 1442, reproducing a design of 1074, is supported by four dragons in the traditionally built courtyard of the old Beijing Observatory. Several other instruments for observing the positions of the Sun, stars and planets are mounted on a high observing platform nearby, with a less obstructed view of the sky.*

ASTRONOMY IN CHINA

Astronomy was practised at the ancient Chinese imperial courts both to forecast the future by observing omens, and to keep the calendar. There is a particularly fine collection of large, early pre-telescopic astronomical instruments at the historic Beijing Observatory, established in 1442 and rebuilt in 1673. Some are replicas of earlier instruments, and others were designed by one of the Observatory's directors, the Jesuit Father Ferdinand Verbiest (1623–88). The instruments are finished in the Chinese style, like this armillary sphere in the Beijing Observatory courtyard (left). Armilliary spheres mimic the orientation of the celestial sphere. They were used to make astronomical calculations and to measure the positions of the Sun and stars across sights sliding on the circles.

A framed quadrant or sextant to measure the angle between two stars had two sighted sticks mounted on the same pivot. Their positions could be oriented along a scale that formed an arc (whose size was one quarter or one-sixth of a circle, hence the name of the instrument) that lay along the line joining the stars. Two people made the sightings on the two stars and co-ordinated their measurements of the positions of the sticks on the scale.

The mural quadrant was a way to measure the angle of stars above the horizon with a large scale which was mounted on a vertical wall aligned north-south. The pinnacle of naked-eye instrument-making on an architectural scale is represented by the Jantar Mantar, an observatory constructed between 1727 and 1734 by Maharajah Jai Singh II of Jaipur, India. It was used to measure the position of the Sun throughout the day and the year.

An astrolabe was a portable instrument used in much the same way as a mural quadrant, with the metal disk and scale (or mater: "mother") of the astrolabe hanging vertically from a ring. The pointing stick, called an alidade, was pivoted at the centre of the disk. A metal map of the sky, called the rete, showing the positions of key navigational stars, could be fitted into a recess on the other side of the astrolabe, together with engraved scales computed for different latitudes. If a navigator made his measurements of the altitudes of several stars fit the scales, this served as a way to calculate latitude and time – the astrolabe was an analogue computer, the prestige satellite navigation device of its time. Astrolabes were invented in the Greek world, possibly by the Greek astronomer Hipparchus (c. 150 BC) and developed into beautiful instruments by Islamic scholars, who broadened their use to calculate the correct time for prayers through the day and to find the qiblah (the direction to Mecca, the orientation for prayer).

Astrology and medicine

In Hellenistic, Indian and Arabic cultures and in Europe up to the second half of the seventeenth century, astrolabes were also used for medical diagnosis and treatment. A doctor could input the latitude and time and use the astrolabe to calculate the appearance of the sky, say at the time of someone's birth. This horoscope could be interpreted in relation to the astrological signs that were supposed to rule the various organs of the body. (No doubt also the bedside hocus pocus with a complicated-looking instrument had a placebo effect, boosting the patient's confidence in the doctor).

Above *An astrolabe made in the mid-sixteenth century by the scientific instrument-maker Gualterus Arsenius in Louvain, Belgium. The open network is a map of prominent stars, which lie at the ends of metal "hooks". The map can be rotated and the stars, measured by the sighting stick, positioned on the contours of angle on the sky engraved on the metal plate behind.*

Right *One of the world's largest sundials at Jantar Mantar Observatory in Jaipur, India. From the shadow of the sloping staircase on the circular scale the Sun's angle can be read and the time told to an accuracy, potentially, of a few seconds.*

Patterns in the sky – the beginning of astronomy

The constellations

The collection of constellations as we now know them was first brought together by Eudoxus of Cnidus (c. 410–c. 350 BC) from research among old manuscripts in the Library at Alexandria. The originals and his own work are lost but there is a poetic account of his treatise called *Phaenomena* by Aratus (c. 310–240 BC). The constellations in his account – which are the northern ones – are mainly people and animals described in terms of Greek mythology.

The earliest depiction of these constellations is a 2000-year old sculpture, known as the *Farnese Atlas*. There is a gap in the constellation figures near the celestial South Pole, which is the region of sky invisible from northern latitudes. From the size and position of the empty circular patch, it appears that the northern constellations originated as a set somewhere at latitude 33° in approximately 1100 BC. This corresponds to the time of the Middle-Assyrian kingdoms. An even earlier indication of knowledge of the constellations comes from early, fragmentary records of a few of them on clay tablets from Mesopotamia dating to about 1700 BC.

The southern constellation figures that now fill the empty patch are modern inventions, dating from the European exploration of the southern hemisphere and the first views of the southern stars. They include figures representing the modern instruments of the day, like the Air Pump, and a symbol of Christian colonialism, the Southern Cross. The latter constellation is much loved in southern hemisphere countries and figures on several national flags.

The earliest representations of individual star patterns are cave paintings in Lascaux, France, and the Cueva de El Castillo in the Spanish mountains, dating to 16,500 years ago, which show patterns of dots in the shape of the asterisms (a recognizable and recognized part of a larger constellation) called the Pleiades, the Northern Crown and the Summer Triangle.

The constellation Ursa Major, the Great Bear, has seven prominent stars in an asterism now called the Plough or the Big Dipper. The four stars in the box shape at the end are the bear, chased by three hunters – the other three stars in a row. In Europe, the Greeks, Basques and Hebrews used this representation, as did cultures in central Asia, but so also did the Cherokee, Algonquin, Zuni, Tlingit and Iroquois nations of North America. Since Native Americans originated by

Below *In 1668 Johannes Vermeer completed* The Astronomer, *modelled on his neighbour in Delft, the optician Anthony van Leeuwenhoek. The astronomer references a book on navigation against a celestial globe by Jodocus Hondius, showing ornate constellation figures of the Great Bear, Draco, Hercules and Lyra.*

Right *The Titan, Atlas, staggers under the weight of the celestial sphere in a Roman sculpture known as the* Farnese Atlas, *the Hellenistic original of which dates from 200 BC. It shows the constellations described by the Greek poet Aratus, and is the oldest surviving depiction of the modern set of constellations.*

migration from the Old World across the Bering Land Bridge about 14,000 years ago , before the Bering Strait had formed, the Ursa Major constellation appears to be at least that old. It may be the oldest cultural artefact still in common use.

Other cultures and civilizations have their own constellation figures. The Chinese constellations are also ancient and include the Big Dipper or Plough, which is represented as an arrangement of clam shells in a mural in a Neolithic grave in Henan Puyang, dating from c. 4000 BC. The same constellation is shown on the Dunhuang star chart, the oldest surviving star chart on paper, dating from about AD 650–680.

Astrology and the planets

Astrology is the belief that the configuration of the planets against the constellations and the zodiacal signs affects what happens on Earth. The personality and future of someone could be predicted from a horoscope, the arrangement of the skies at the moment of birth or conception (this is the basis for the crude horoscopes published in many newspapers); or the outcome of a venture could be predicted from the moment of its inception. Finally, particularly in India, astrologers attempt to answer any question posed by looking at a horoscope cast at the moment the question was asked. There are different rules for relating the astronomy to the predictions. These include Indian and Chinese, systems of astrology, but the most-used system in the western world derives from Greco-Roman Egypt of the first century AD. While astrology is regarded by modern scientists as a superstition, astrology represents the start of astronomy.

The need to calculate planetary positions for a horoscope was the reason that the astronomer Claudius Ptolemaeus (known as Ptolemy, c. AD 90–c.168), a Roman citizen who lived in Alexandria, wrote his astronomical treatise known as the *Almagest* (*The Great Treatise*) and its astrological companion volume known as the *Tetrabiblos* (*Four Books*). The two works have passed to us through Arabic editions. They put forward a theory of the Universe in which the planets were mounted on nested crystal spheres that rotated around the Earth, which was therefore at the centre of the Universe, and showed how to calculate the positions of the planets.

Opposite *Portrait of Ptolemy by Joos van Ghent, c.1475. Ptolemy is crowned, confusing the second century astronomer with the dynasty of Egyptian kings.*

Above *The constellation of the Great Bear from* Urania's Mirror, *a set of hand-coloured cards by "a lady" (in reality the Reverend Richard Rouse Bloxam, London, 1825). The stars are perforated with little holes, to suggest the constellation at night if the cards are held up to a candle.*

Ideally, as the Greek philosopher Aristotle (384–322 BC) had proposed, the celestial bodies should move in perfect circles. In fact, the planets move sometimes faster and sometimes slower; they even, in the case of Mars especially, slip back against the background of fixed stars (retrograde motion). Ptolemy developed a complex system of "epicycles" to explain this. In its simplest form, the planet moves on a circle whose centre moves on a circle around the Earth. In fact, dozens of epicycles (some say as many as 80) are needed to provide a satisfactory representation of the motions of the planets, and the number and complexity of the system of epicycles has made them a metaphor for bad science.

Below *The circumpolar constellations, including (bottom) the Plough/Big Dipper, on the Dunhuang star chart (AD 650–680). The chart was discovered in 1900, bricked up since about AD 1000 in a Buddhist temple in western China in a library of over 30,000 volumes, where it had been hidden from a marauding gang.*

ZODIAC

The Zodiac is the track of the Sun and planets through the sky; the concept dates back to Babylonian astronomy. Zodiac means "to do with animals", referring to the original six constellations along the track, which had the shape of animals. At a later date, non-animal examples like Libra (The Scales) were interpolated to make 12 zodiacal constellations. The circle that defines the exact path of the Sun is now divided into 12 equal divisions called the Signs of the Zodiac; these are distinct from the constellations, but each given the name of the constellation that they originally spanned (although the Signs and the constellations have slipped apart over the years).

Right *The planets orbit the Earth in a plane, and are always seen projected against the band of the Zodiac and its constellations and Signs. This perspective drawing of the solar system is from an engraving in* Atlas Coelestis *by Andreas Cellarius, Amsterdam, 1660, and is described by him in the corner as Tycho's model. In fact, it has Mercury and Venus orbiting the Sun, and the Sun and all the other planets orbiting the Earth, and is a little-known progenitor of Tycho's model due to the Roman intellectual, Martianus Capella (c. fifth century AD).*

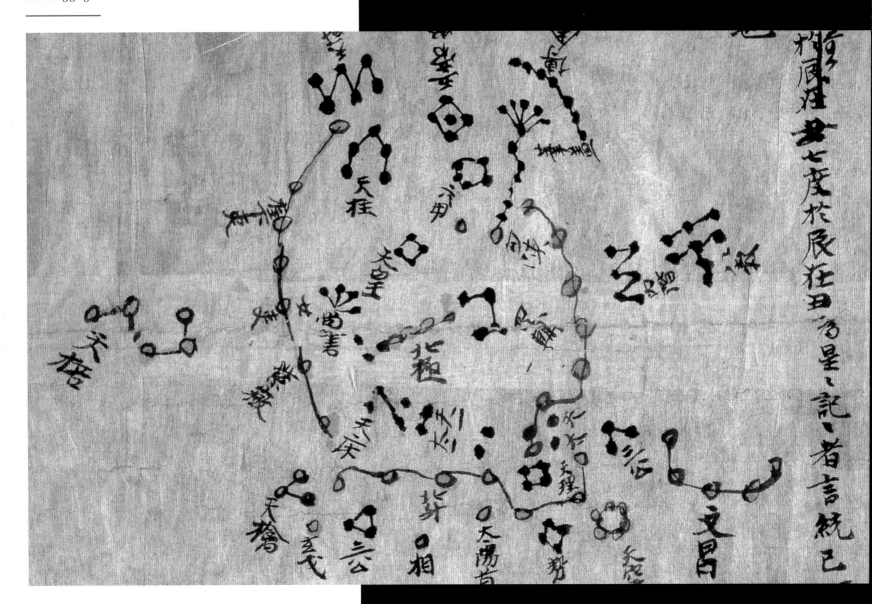

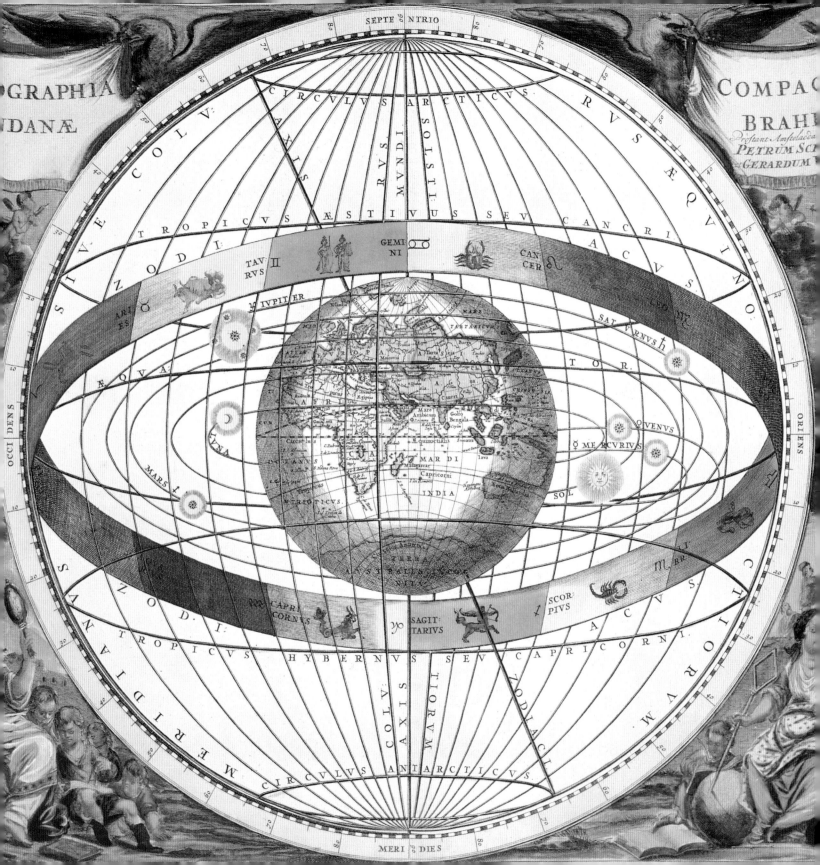

4
The nature of the Earth-centred universe

Thales of Miletus (c. 624–c. 546 BC) is called the first of the Greek philosophers. According to the historian Herodotus – in what is a doubtful story – he predicted the famous solar eclipse of 585 BC. This eclipse occurred during a battle between the armies of Lydia and the Medes, who became so concerned about the portentous celestial omen that they threw down their weapons and made peace.

Thales put forward a theory that everything was made of water and the phenomena that we see are due to the water's changing forms. This was the first theory in the physical sciences that attempted to break down phenomena into simpler elements. This theme was developed by Aristotle (384–322 BC), undoubtedly the most important Greek philosopher. He integrated his ideas about the structure of matter into an overall theory of the structure of the Universe. Cosmologists have forever afterwards been attempting to do the same.

Aristotle thought that the elements of which the Earth is composed are different from those in the heavens. The four elements that make up everything mundane, or terrestrial, are earth, air, fire and water. Iron, for example, is primarily made of the element earth, with small amounts of the other three. The heavens are made of a fifth element called "aether" or "quintessence" (meaning fifth element), which is weightless and which never changes.

According to Aristotle, the four terrestrial elements tend to move toward their natural place – air upwards towards the sky and earth downwards towards the centre of the Universe. The Earth is at the centre, surrounded by a series of concentric crystal spheres carrying the Sun, Moon, planets and stars. They move eternally around the Earth in an unchanging circular motion. The motion has to be basically circular because the celestial bodies are perfect and the circle is the perfect geometric figure. To account for the observed non-uniform motion of the planets, Aristotle suggested that the spheres that carry the planets are embedded within other spheres, perhaps 50 of them altogether. Each sphere is moved by an unmoved mover. There is a "prime mover" for the outermost sphere of fixed stars. Aristotle did not think of the prime mover as a physical entity that pushed the outer sphere, but other more literal-minded people did. In a famous woodcut, drawn in a fake medieval German style, the nineteenth-century French astronomer Camille Flammarion (1842–1925) depicted the prime mover as a mechanism with gears and wheels moving the outer sphere.

Below *Aristotle's bronze portrait bust by the sculptor Lysippus (both men were employed at the same time by Alexander the Great) is known from contemporary copies like this one in marble, although the original has been lost. The features here are the authentic face of the Greek philosopher aged about 40.*

Opposite above *A pilgrim travels to the edge of the world and looks through the celestial sphere to the mechanism beyond that turns the stars and planets. The picture from 1888, in the style of a German woodcut, is the invention of the French astronomer and populariser Camille Flammarion but nicely illustrates the Ptolemaic cosmology of medieval times.*

THE FIRST COMPLEX COMPUTER

Flammarion's imaginative representation of the mechanism of the Universe was given a remarkable physical form in a clockwork-like calculator known as the Antikythera mechanism, a complex geared calculator found in a shipwreck off the Greek island of Antikythera and dating to about 150–100 BC. It was built according to the theories of the Greek astronomer Hipparchus (c. 190–c. 120 BC). He compiled an accurate star catalogue to make it possible to identify any new stars that appeared. Comparing his measurements of star positions to previous measurements he discovered precession, the wobbling of the Earth that changes the apparent positions of the stars as they rise and set across the sky. Using data that he found recorded by earlier Babylonian astronomers, he developed geometric constructions to calculate the positions of the Sun, the Moon and the planets. These geometric constructions are the basis of the Antikythera mechanism.

Left Corroded by 2000 years under the Mediterranean Sea off Crete, the Antikythera Mechanism shows some of its over 30 gears.

SHAPE OF THE EARTH

The geographer Eratosthenes of Cyrene (c. 276–c. 195 BC) was a librarian at the Library in Alexandria and investigated the shape of the Earth. Although some early philosophers thought the world was flat, like Leucippus (c. 5th century) and Democritus (c. 460–370 BC), there were well-rehearsed arguments that it was spherical. Two of them were that the shadow of the Earth cast on the Moon during a lunar eclipse was always circular and must be the shadow of a spherical body, and that a lookout at the top of a ship's mast would see land before his fellow sailors on deck, because he was able to see over the curvature of the Earth. Eratosthenes determined the size of the Earth. He had heard that at Syene in Upper Egypt (present-day Aswan) the Sun was directly overhead at noon on the day of the summer solstice — the Sun's rays reached the bottom of a deep well. He determined the length of the shadow of a vertical post at Alexandria on the same day and found that the angle of the Sun was one fiftieth of a circle to the south of the zenith. He calculated the distance between Alexandria and Syene by driving a carriage between the two cities and counting the revolutions of the wheels. The circumference of the Earth was 50 times this, or 250,000 stadia. The equivalent of the length in stadia is not well known but is believed to be about 45,000 kilometres (28,000 miles), remarkably close to the modern measurement of 40,000 kilometres (25,000 miles). It is surprising that, although most educated people knew that the world was spherical, the idea that it was flat persisted in common thought. When Columbus set out in 1492 on the voyage from Spain to the Americas, it was popularly speculated that his ship would fall off the edge of the flat, disc-shaped Earth, but Columbus knew that this was fanciful and the dangers he faced were practical — storms, mutiny, inadequacy of supplies and possibly sea monsters.

Top left *California artist Antar Dayal's picture (c. 2000) of the pre-Columbian notion of a flat Earth, ships plunging off the edge towards dangerous monsters.*

Above *A time-lapse montage of the Moon during the total lunar eclipse of 21 January 2000. The shadow of the Earth cast on the Moon during an eclipse always has a circular edge, showing that the Earth is spherical. The orange-red colour of totality is due to light refracted in the Earth's atmosphere onto the lunar surface.*

Below *Map of the world as described by Eratosthenes, showing the land centred on Alexandria where he worked in the Library.*

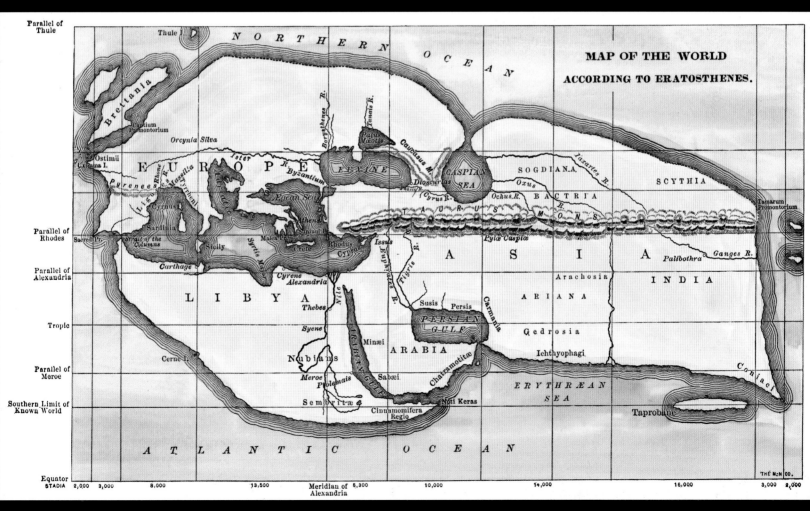

Astronomy in the dark

Greek civilization gave way to Roman rule in the second century BC, but Rome adopted Greek culture, so that Roman scholars like Ptolemy who wrote scientific treatises did so in Greek. After the fall in the west of the Roman Empire in the second half of the fifth century, astronomy developed separately in medieval western Europe and in the Islamic world.

Astronomy and the Church

After the collapse of Roman rule, Europe entered the period sometimes known as the "Dark Ages" in which knowledge of many Roman advances was lost and during which astronomical theories were adapted to Christian thought. The original work that was accomplished in the field concentrated on two aspects: timekeeping, which was necessary to be able to celebrate services and festivals at appropriate times in unison across Christendom, and cosmology, which made science consistent with theology.

The synthesis of Aristotelian cosmology with Christian theology took some time. There was no obvious connection between what Aristotle thought and the writings of the Old Testament. This link was forged in the thirteenth century by the Thomists, intellectuals at the University of Paris, led by a Dominican friar Thomas Aquinas (1225–74), later canonized. Aristotle's Prime Mover was identified as God, who had assigned an angel to each of the eight celestial spheres to move them individually. Outside the celestial spheres of the seven planets and the "firmament" of fixed stars was thought to lie within a ninth sphere, identified with "the waters above the firmament". Outside everything lay the motionless Empyreum, the dwelling place of God – which the Thomists identified with heaven. God is perfect, and therefore the heavens must be perfect and eternal, distinct from our own existence on Earth. This philosophy became the accepted teaching of the established Church.

Timekeeping was perhaps more mathematical in nature and rather intricate, but in principle easier and less controversial. Teachers at monasteries and universities across Europe, like the Irish monk Johannes de Sacrobosco (John of Holywood, c. 1195–c. 1256), who taught at the University of Paris in the thirteenth century, developed textbooks which taught mathematical astronomy, including "dialling" – the construction and use of sundials for telling time during the day. To tell the time at night, navigators and astronomers had the nocturnal, an instrument first described by the Spanish navigator Martín Cortés de Albacar (1510–82) in 1551. It used the Pointers of the constellation of the Great Bear as the hour hand of a 24-hour clock, pivoted around Polaris.

Left *A portable brass combination Sundial and Nocturnal, made by Egide Coignet (Antwerp, c.1560). To tell the time at night, set the rotating scale so that midnight is lined up on the date (about 7 May in this example), hang the instrument by the hole at the top, sight Polaris through the hole in the middle and rotate the indicator to align with the two stars of the Pointers of the Plough/Big Dipper. The indicator shows the time, in this case, unrealistically, 3 o'clock in the afternoon, in daylight.*

Right *Hans Holbein (the younger) painted* The Ambassadors *in 1533, presenting the portrait of two French officials at the London court of Henry VIII, rich and educated men with a sophisticated knowledge of the arts and sciences, including astronomy, which is represented by the dazzling array of instruments on the table, such as the sundial and the celestial sphere.*

    The calendar posed a significant problem. There are marginally less than 365.25 days in a year, so that if you wish to keep the calendar date in line with the cycle of the seasons you have to account for the extra day in the year that occurs every four years. To do this, the Julian calendar, introduced in Rome by Julius Caesar in 45 BC, used a year of 365 days, with a leap day added every fourth year to make an average of 365.25 days. Even in this formulation, however, the seasons gradually changed their position in the calendar because there are actually 365.24219 days in a year, and by the sixteenth century the seasons were more than a week adrift. This problem was solved for the western Christian churches by the reform of the calendar under the authority of a papal bull issued by Pope Gregory XIII in 1582, by which ten days were omitted from October that year, and a twist was added to the system of leap days by omitting them from three century years every four. It was, however, literally centuries before the Gregorian calendar became widely accepted across Europe.

    The supreme calendrical problem in Europe was the definition of the date of Easter. Easter commemorates the events of the Crucifixion, whose time was set with respect to the Jewish festival of Passover. The Jewish calendar is regulated by both the annual cycle of the Sun and the monthly cycle of the Moon, with Passover set at the time of the first Full Moon after the spring equinox. Due to uncertainties in predicting the motion of the Sun, but also the even greater uncertainties of the Moon, the churches of Christendom came to celebrate Easter at different times. This was also cleared up – for western churches – at the same conference in 1582, but even in the twenty-first century not all Eastern Churches accept the sixteenth-century reforms in relation to their celebration of Easter.

Above *Pope Gregory XIII chairs the commission for reforming the Roman Calendar (1582–83), and is lectured by an astronomer on the details of the slippage of the zodiacal signs over time.*

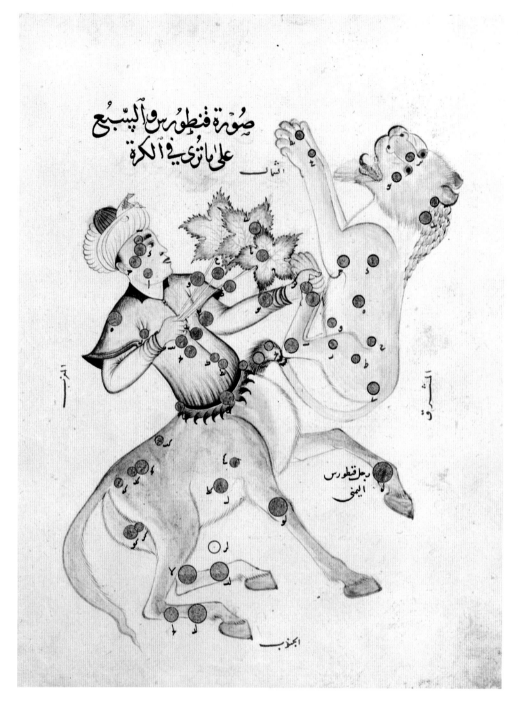

Islamic astronomy

Within a century of the death of the Prophet Mohammed (in the year 632 of the Christian Era), Baghdad had become the de facto capital of the Islamic world and the caliphs Harun al-Rashid (763–809) and al-Ma'mun (786–833), established a library and a translation centre to render Greek textbooks into Arabic. This created the foundation of Islamic science, in which astronomy was both developed as an end in itself and as an aid for correct Islamic practice, for example in calculating the position of the qiblah (direction to Mecca) and determining the correct times of the day to pray. The Arabic translations of ancient Greek works are sometimes the only versions that have survived to the present day, occasionally in revised versions such as the revision of Ptolemy's star catalogue by Abd al-Rahman al-Sufi (903–986), with its wonderful constellation figures.

Among the observatories that were founded to determine new, more accurate data for the calculations was the one in Samarkand (in present-day Uzbekistan), established by Ulugh Beg (1394–1449), its provincial governor and later ruler. Ulugh Beg himself was the principal astronomer there and supervised the construction of a gigantic sextant to measure the positions of the stars. Other astronomers concentrated on the preparation of collections of tables, called a zij, to facilitate the calculation of the positions of the planets. Yet others developed Aristotelian cosmology, like Muhammed ibn Rushd (known in the west as Averroes, the Commentator (1126–98)).

From about the year 700 of the Christian Era to about 1500, Islamic astronomy and other sciences flourished, preserving and developing for about a millennium foundations that had been established in works of classical antiquity that otherwise would have stagnated in Europe, or would even have been lost. The legacy of this period can be seen in in astronomical words derived from Arabic that are used in European languages in the present day; words like cipher and algorithm (a mathematical formula), azimuth, zenith and nadir, almanac, and the names of stars like Altair.

Above *The constellations of Centaurus and Leo from a fifteenth century copy of Al Sufi's star catalogue.*

Right *Assistants work at the observatory of the Islamic astronomer Taqi al-Din ibn Maruf at Istanbul, founded in 1577 (that is the boss, the large figure in black, kneeling in the right background). The miniature painting is from the "History of the King of Kings", an epic poem by Ala ad-Din Mansur-Shirazi, c. 1581.*

6

Revolutions of the heavenly spheres: Copernicus

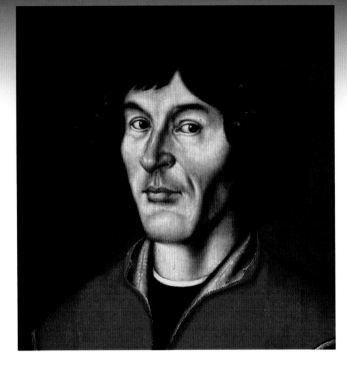

Right *Portrait of Nicolaus Copernicus from Toruń, one of the two earliest portraits, perhaps painted from life in the early sixteenth century. Aged about 40, he is dressed unpretentiously in a red sleeveless tunic, which is however lined with fur.*

Niclas Koppernigk (known as Nicolaus Copernicus, 1473–1543) was born in Toruń in Poland; he was educated in Cracow in Poland and in Bologna and Ferrara in Italy. In Italy, he developed an interest in astronomy and medicine, with which astronomy was then thought to be connected. In 1500, Copernicus lectured on astronomy in Rome. He returned to Poland in 1503 as a canon in the cathedral of Frombork, a position he retained until his death.

Teachers often remark that they never really understood a subject until they came to teach it, and so it must have been for Copernicus. In teaching the motions of the planets according to Ptolemy's descriptions of epicycles, Copernicus became dissatisfied with what he was saying. Some of the geometric constructions necessary to the epicyclic calculations seemed very arbitrary. And why did the period of the Sun around the Earth (the year) find its way into the epicyclic motions of every planet? – "It is clear that each of the six planets in its motion shares something with the Sun, and the Sun's motion is so to speak the mirror and measure for their motions," wrote the Austrian astronomer Georg von Peurbach (1423–61), and Copernicus agreed.

In Frombork, Copernicus circulated a manuscript showing that the feature of the planetary orbits that Peurbach had highlighted would be a natural consequence of a solar system in which the Sun was the centre – naturally, if the Earth was in orbit around the Sun, as seen from the Earth each planet would seem to revolve around the Earth in a motion that would be the result of the addition of its own motion around the Sun and the Earth's motion around the Sun. At the suggestion of a pupil, the Austrian astronomer Georg Joachim von Lauchen (known as Rheticus, 1514–74), Copernicus worked up a small publication on this topic. It proved uncontroversial and Rheticus persuaded Copernicus to compile a major work, which Rheticus saw through the press as *De Revolutionibus Orbium Coelestium* (*On the Revolutions of the Celestial Spheres*), published in Nuremberg in 1543. In the book, Copernicus put forward the concept that the planets revolved around the Sun, in outwards order: Mercury, Venus, Earth, Mars, Jupiter and Saturn; while the Moon revolved around the Earth. The book is regarded as the foundation of the heliocentric (Sun-centred) theory of the solar system. Copernicus received a copy of the book, hot off the press, only on his deathbed.

In an attempt to deflect criticism, Andreas Osiander included in the published work a preface by himself (but unsigned and in consequence appearing to be by Copernicus), in which it was stated that the author was not asserting that the Earth actually moved round the Sun, only that this was a convenient way of making the calculations. Copernicus showed that the puzzling retrograde motions of the outer planets, particularly Mars, were a natural consequence of the way that the inner planets revolved around the Sun more quickly than the outer ones – an athlete running quickly on the inside track of a racecourse would see an athlete in front on an outer track moving ahead, but then as he overtook he would see him apparently falling behind.

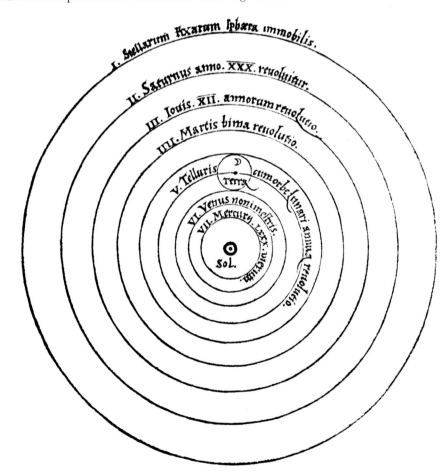

Below *Copernicus' own diagram of the solar system from a copy of his book,* De Revolutionibus Orbium Coelestium *of 1543, with the Sun (Sol) in the middle of the orbit of the planets, the Earth going around the Sun "in its annual revolution, with the orbit of the Moon".*

Most of the book was dauntingly technical. Copies were pored over and annotated by scholars, but its main hypothesis that the Earth revolved around the Sun at first attracted little controversy. It was denounced in 1546, soon after its publication, as being contrary to Scripture by a fundamentalist Dominican, Giovanni Maria Tolosani (c. 1471–1549), but it was not until Galileo proved Copernicus's heliocentric theory of the solar system by his observations with the telescope in 1610 that the affair took on a very high profile.

Left *Rheticus shows Copernicus, on his deathbed, a copy of his book* On the Revolutions of the Celestial Spheres, *hot off the press.*

Below *The orbit of Mars from the year 1580 (beginning right of centre) to 1596 (ending upper left) according to Ptolemy's model of the solar system, with the Earth stationary in the centre of the complex path. While Mars travels mostly anticlockwise, it seems to go backwards (retrograde motion) at the loops in the orbit. This figure is from Kepler's book of 1609 on the orbit of Mars,* Astronomia Nova, *based on measurements by Tycho Brahe. The book showed how simple and appealing Kepler's theory of the motion of Mars was, compared to Ptolemy's, as shown here.*

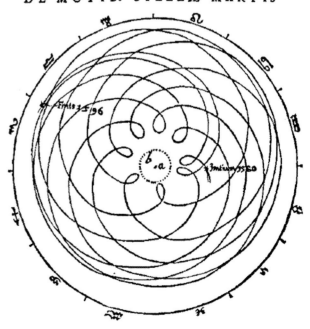

Measuring the skies from the City of the Stars

To test whether his heliocentric theory of the Solar System was better than Ptolemy's geocentric one, Copernicus had to rely on old measurements of the positions of the planets. In the last half of the sixteenth century, a Danish astronomer, Tyge Ottesen Brahe de Knudstrup (known as Tycho Brahe, 1546–1601), compiled much more accurate measurements.

Tycho had become interested in astronomy at the age of 14, and as a student noticed discrepancies in the tables of planetary positions in his textbooks. He set out to make the most accurate measurements he could. He was spurred on by the appearance in 1572 of a new star, whose position he measured assiduously, proving that it was indeed a star and not some atmospheric phenomenon. His publication on the subject brought him fame and King Frederick II of Denmark became his patron, providing him with the island of Hven and an observatory there, Uraniborg (Castle of the Heavens), later Stjerneborg (Castle of the Stars). In 1588, the king died and the throne passed to his 10-year old son Christian. While King Christian IV was still a child, Tycho's friends protected his interests at court, but as Christian grew up he showed less of an interest in astronomy than his father and reduced support for Tycho's observatory. Tycho spent his final years in Prague, under the patronage of Emperor Rudolf.

Tycho's invaluable legacy to science was his collection of observations but he also left a theory of the Solar System that he had developed with the aid of those observations. It was a compromise between Ptolemy's geocentric theory and the Copernican heliocentric model. The Earth was at rest in the centre and the Moon and Sun revolved around it; the other planets – Mercury, Venus, Mars, Jupiter and Saturn – revolved around the sun. In the event, the theory just complicated the issue and did not catch on.

Right *In an unsuccessful and confusing compromise between Ptolemy's and Copernicus' theories of the universe, Tycho's model of the Solar System has some planets orbiting the Earth, some the Sun. Engraving from* Harmonica Macrocosmica *by Andreas Cellarius, Amsterdam, 1708.*

Right *Tycho Brahe dressed in the finery of a Danish nobleman in the only known authentic portrait (a contemporary copy of a lost portrait painted from life).*

TYCHO BRAHE

Tycho was a Danish nobleman, rich enough to indulge his appetites and his interests (he was reckoned at one time to own one per cent of the wealth of the country). There is a multitude of stories about his eccentricities. He had an artificial nose of gold, having lost part of his actual nose in a duel. He had a pet elk that died after falling, drunk, downstairs. Tycho himself died after a formal dinner from retention of urine, having been too embarrassed to excuse himself from the table. His monument in the Church of Our Lady of Týn, in the Old Town Square in Prague shows him portly in armour, with jowls and a large moustache.

ORTHOGRAPHIA PRÆCIPVÆ DOMVS ARCIS VRANIBVRGI in Insula Porthmi Danici Venusia, *vulgo* Huenna, Astronomiæ instaurandæ gratia, circa annum MDLXXX, à TYCHONE BRAHE exædificatæ.

URANIBORG AND STJERNBORG

Tycho's observatory had the most modern laboratories and instruments, including an enormous Great Mural Quadrant. It also had a library and platforms on which to erect portable instruments for making astronomical observations. There were three rooms for visiting scientists and dignitaries, and a number of smaller rooms for assistants. The more-than-100 assistants came from universities all over Europe and worked in return for free food and lodging. Tycho was very particular to maintain his ownership of the astronomical data and the assistants had to sign confidentiality agreements. The observatory also hosted other scientific studies in chemistry, medicine, horticultural refinement, meteorology and cartography. Uraniborg was the first observatory in the modern style.

Above *Brahe's first scientific laboratory, Uraniborg, was a Disneyesque castle, and proved unusable as an observatory since the instruments were exposed on towers and shook in the often tempestuous winds off the North Sea. Its cramped architecture could not house Brahe's ambitiously large new instruments.*

Right *Stjerneborg was built underground with the astronomical instruments deployed on platforms at ground level behind wind-breaks.*

8

Galileo Galilei

Galileo Galilei (1564–1642) is likely to figure on every scientist's list of their ten favourite scientists – perhaps even at the top – partly because of the range of his work and in part because of the stance he took against authority, the same principle that is epitomized in the motto of the Royal Society of London, *Nullius in verba* – "Take nobody's word for it". Nevertheless, this position was compromised by Galileo's pragmatic surrender to the Inquisition in Rome when his work brought him into conflict with the Church.

In 1592, Galileo became a professor at the University of Padua, where in 1609 he heard of the invention in the Netherlands of the telescope by Hans Lipperhey (c. 1570–1619), a spectacle maker in Middelburg. It is said that, while playing in Lipperhey's shop, his grandchildren noticed that on holding two particular lenses up, one in front of the other, they could see the rigging of ships in the harbour more clearly. Lipperhey tried unsuccessfully to patent his discovery and the news got out.

Galileo made a telescope in 1609, which magnified about three times (eight times in area), improving on this later the same year with one that magnified 20 times. Galileo offered a telescope to the city of Venice and took some officials to the top of a tower from which, with the aid of the telescope, they could identify ships approaching the harbour that were invisible without it. They awarded Galileo an increase in salary.

While not the first to do so – that honour goes to map-maker, New World colonist and mathematician Thomas Harriot (1560–1621), who published none of his work – Galileo turned a telescope to the study of astronomy in November and December 1609. He understood what he saw and that it amounted to a revolution in astronomy.

With his telescope, Galileo saw that the Moon has mountains and dark flat areas – Galileo thought them to be seas, but they are dusty plains – just like those on Earth. The Moon has no light of its own, but reflects that of the Sun. In the unilluminated parts, Galileo saw bright spots, manifestly mountain-tops that were catching sunlight before the lower valleys. His observations of the Moon showed that it was rough, indicating that the heavens are not perfect. This was confirmed by his observation of spots on the Sun. Moreover, the celestial bodies are not different from the Earth (the Moon as seen from the Earth looks like how Galileo could imagine the Earth would look from the Moon).

Early in January 1610, Galileo observed the planet Jupiter. He saw what he thought were three fixed stars near it, strung out on a line through the planet. The following night he saw the three stars on the other side of Jupiter. Over the next week he saw that the little stars never left Jupiter but changed their position with respect to each other and the planet – in fact there were actually four of these little stars oscillating from side to side. How did they move through each other and through Jupiter? Ten days later he had a flash of insight: the little stars must be four moons, which revolve around Jupiter. Jupiter is the centre of a mini-solar system. It all added weight to the Copernican model of the solar system.

Galileo put the moons of Jupiter to practical use: he named them the Medicean stars after Cosimo II, the head of the famous Medici family in Florence. Galileo had tutored Cosimo in mathematics as a young man and was seeking patronage. Cosimo offered Galileo a court position.

Left *Galileo Galilei, painted from life by Justus Sustermans, from Antwerp, portraitist to the Medici family. It is one of the last portraits of the aging astronomer and shows him with his telescope.*

Right *Two Galilean telescopes (possibly made by Galileo himself) mounted on a presentation stand with (at the centre of the oval, ivory mounting), the lens with which he discovered the satellites of Jupiter. Galileo gave the lens to Grand Duke Ferdinand II; it is now cracked after an early careless inspection.*

Right *In 1609, Galileo drafted a letter to Leonardo Donato, Doge of Venice, offering him one of the telescopes that he had made, and pointing out its potential use for the defence of the city against invading fleets of ships. He sent the fair copy of the letter on August 24, 1609, keeping the draft (shown here) as a copy. In January 1610, he used the blank space at the bottom of the sheet for notes about his first observations of the planet Jupiter and four of Jupiter's moons.*

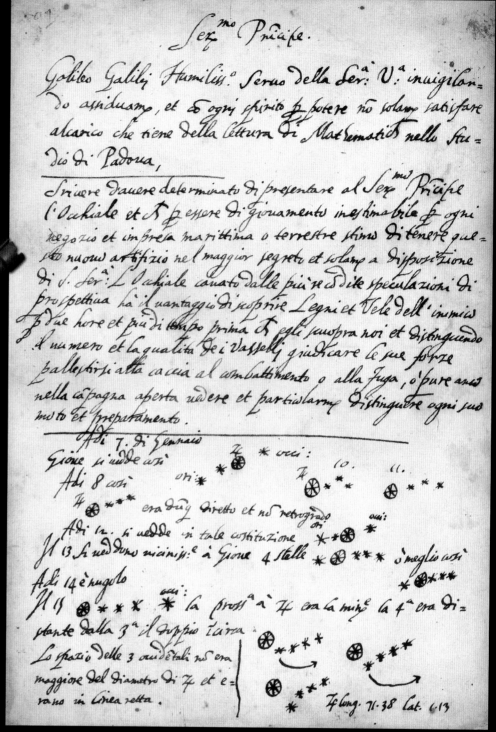

Translation:

Most Serene Prince.

Galileo Galilei most humbly prostrates himself before Your Highness, watching carefully, and with all spirit of willingness, not only to satisfy what concerns the reading of mathematics in the study of Padua, but to write of having decided to present to Your Highness a telescope that will be a great help in maritime and land enterprises. I assure you I shall keep this new invention a great secret and show it only to Your Highness. The telescope was made for the most accurate study of distances. This telescope has the advantage of discovering the ships of the enemy two hours before they can be seen with the natural vision and to distinguish the number and quality of the ships and to judge their strength and be ready to chase them, to fight them, or to flee from them; or, in the open country to see all details and to distinguish every movement and preparation.

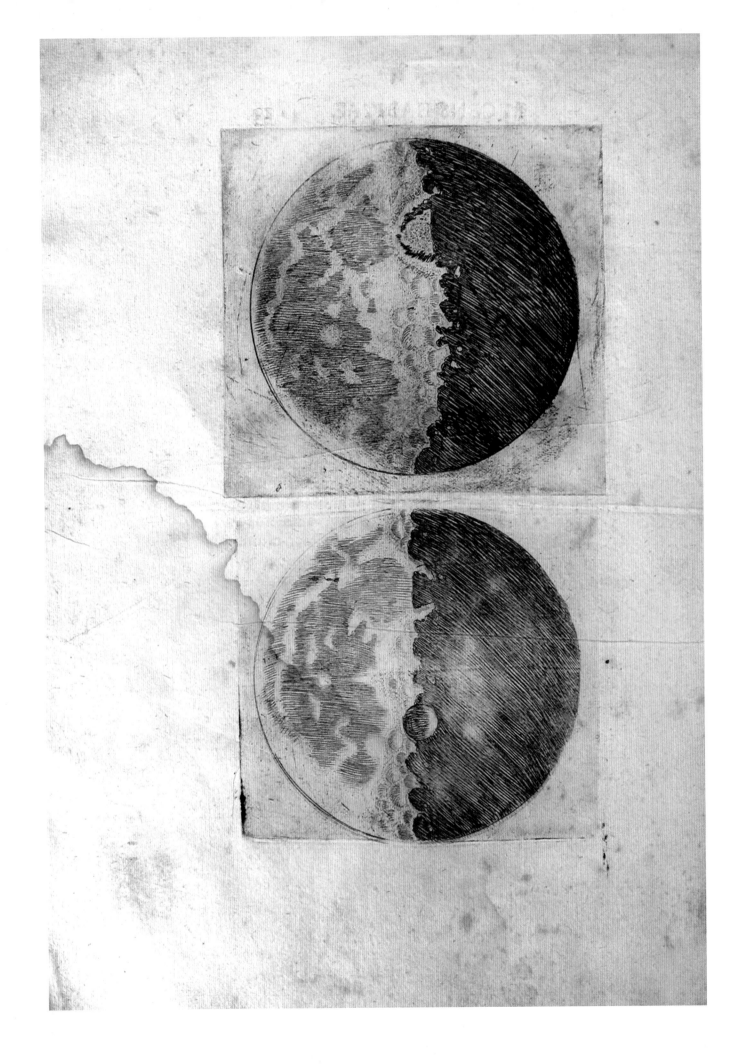

Opposite *Galileo's first four images of the Moon were published in 1610 as engravings in his book* Sidereus Nuncius (Starry Messenger), *two of them reproduced here. Galileo's telescope had a field of view much smaller than the disc of the Moon so he must have made sketches of what he could see smaller bits and assembled them later into one picture like a jigsaw. As a result his pictures do not exactly correspond with modern photographs and accurate maps, and it is surprisingly hard to identify particular features, like the prominent circular crater seen in the lower picture on the "terminator" or line between lunar night and day. It is generally thought to be the crater Albategnius but it is much too large. The pictures do however very accurately correspond in general to what he wrote about the appearance of the Moon:*

On the fourth or fifth day after New Moon, when the Moon presents itself to us with bright horns, the boundary which divides the part in shadow from the enlightened part does not extend continuously in an ellipse, as would happen in the case of a perfectly spherical body, but it is marked out by an irregular, uneven, and very wavy line, as represented in the figure given, for several bright excrescences, as they may be called, extend beyond the boundary of light and shadow into the dark part, and on the other hand pieces of shadow encroach upon the light : nay, even a great quantity of small blackish spots, altogether separated from the dark part, sprinkle everywhere almost the whole space which is at the time flooded with the Sun's light, with the exception of that part alone which is occupied by the great and ancient spots. I have noticed that the small spots just mentioned have this common characteristic, always and in every case, that they have the dark part towards the Sun's position, and on the side away from the Sun they have brighter boundaries, as if they were crowned with shining summits. Now we have an appearance quite similar on the Earth about sunrise, when we behold the valleys, not yet flooded with light, but the mountains surrounding them on the side opposite to the Sun already ablaze with the splendour of his beams; and just as the shadows in the hollows of the Earth diminish in size as the Sun rises higher, so also these spots on the Moon lose their blackness as the illuminated part grows larger and larger.

Galileo, *Sidereus Nuncius*, trans. Edward Stafford Carlos (1880).

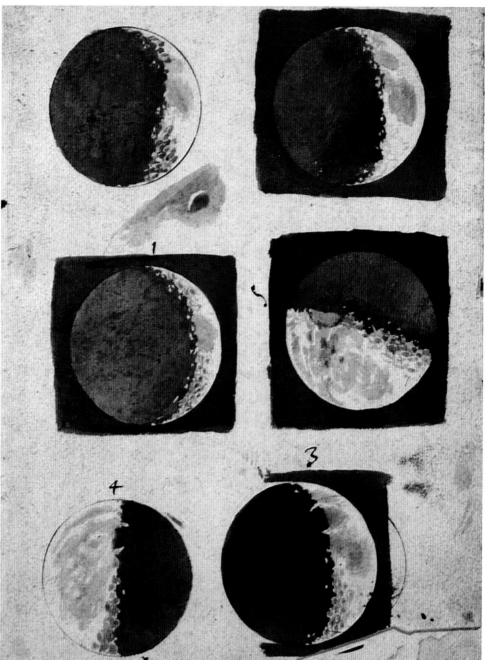

Above *Watercolour sketches of the Moon by Galileo (bound in his manuscript copy of* Siderius Nuncius *now held in the National Library in Florence). The sketches are apparently originals made by Galileo from observations at the end of 1609 or beginning of 1610.*

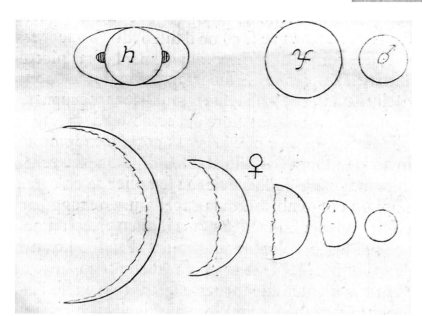

Left *In his book on comets,* Il Saggiatore *(The Assayer, Rome, 1623), Galileo showed an engraving of the planets as seen by him through his telescope. Mars and Jupiter (top right) are simple discs, and Saturn (top left) is complicated. Venus (bottom row) shows phases, from a large, thin crescent to a smaller, full disc. This could only happen if Venus was a solid planet, illuminated by the Sun, in orbit around it, and seen from the Earth from further away.*

GALILEO AND THE CHURCH

At first Galileo's beliefs were regarded by the intellectuals in the Vatican as acceptable scientific advances, as they were by the community of scientists across Europe. Eventually, however, they got him into terrible trouble, although his friends, like Cardinal Bellarmine, were able to protect him from the very worst. He was denounced to the Inquisition, and was subjected to a long process of investigation, eventually being tried for heresy. Under pressure, his protectors had to agree that the Inquisition could show Galileo the instruments of torture if not actually use them. Galileo well understood the implicit threat, capitulated, was convicted and abjured. He was placed under house arrest and told not to teach the Copernican theory, even as a theoretical hypothesis. In 1642, by that time blind, he died at home.

Left Galileo in front of the Holy Office of the Vatican by the French painter Joseph-Nicolas Robert-Fleury (1847). Galileo is being tried in a large room, guarded by a cold and menacing soldier. On the wall is the heavily metaphorical Disputa del Santo Sacramento by Raphael, a cloud layer dividing the eternal certainties of heaven and the Holy Spirit, above, from, below, the disputatious, fallible argument of worldly theologians, heretics and Popes, like the members of the Inquisition seated at the table at the trial. Galileo is self-confident, standing firm with scientific arguments garnered from disorganized books against the unyielding cardinal in front of him. This work is by a romantic painter with a heroic, anti-clerical story to tell that is different from the historic reality of Galileo: an infirm old man, frightened by the threat of torture, kneeling and trembling hands laid on the Holy Scripture humbly recanting.

Right Astronomical Observations: Jupiter *by Bologna artist Donato Creti (1711). Creti was commissioned by Luigi Ferdinando, Count de Marsigli (1658–1730), to make a series of paintings of astronomical subjects in rural landscapes as a present for the Pope Clement XI, to convince him to sponsor an astronomical institute in Bologna. Marsigli's gift and his representations to the Pope succeeded and the observatory was established in 1714, Italy's first public observatory. Creti painted the landscapes in the eight paintings himself and commissioned the miniaturist Raimondo Manzini (1668–1744) to add representations of astronomical phenomena in the sky above, using notes and sketches by the astronomer Eustachio Manfredi (1674–1739) and his own telescopic observations, made under Manfredi's guidance. In the painting called* Jupiter, *two astronomers in the foreground, with a telescope on a stand, discuss their observations. Ranged each side of Jupiter are three of the planet's four moons, and its surface is covered with six major cloud bands and the Great Red Spot.*

Telescopic revelations

From September 1610, Galileo was able to observe the planet Venus. By mid-November, he had observed that it had phases, like the Moon. He announced this in an anagram which he later decoded, a two-stage process that established his priority to the discovery, guarding against the possibility that someone could hear of it and falsely claim it as his own while letters were still in transit to their intended recipients. (The process also hedges bets against some sort of mistake, giving time and scope for second thoughts before the last stage of decoding.) In a letter to the Tuscan ambassador in Prague, Giuliano de Medici, in December 1610 Galileo included the anagram *Haec immatura a me iam frustra leguntur o y*, which is not very good Latin for "This was already tried by me in vain too early."

When the letters are unscrambled, the message is *Cynthiae figuras aemulatur mater amorum* ("The mother of loves [Venus] imitates the shapes of Cynthia [the moon].") Galileo also noted that as Venus changed from a crescent shape to a full round disk it got smaller, so it was receding from the Earth. This showed how Venus's orbit extended behind the Sun. In Ptolemy's theory, Venus moves in an orbit between the Earth and the Sun and could never look like the Full Moon, as a full circle.

With his telescope, Galileo was able to see stars that had been up to then invisible. He could see that the Milky Way might be made up of huge numbers of small stars amassed together to create the milky light. He gave examples of the numbers of new stars that he had seen for the first time in well-known clusters of stars, such as the Pleiades, the Praesepe, and the belt and sword regions of Orion. Whatever the scientific conclusions that could be drawn from this, it was clear at least that the stars were not created for mankind's benefit – there could hardly be any benefit in stars that could not be seen over the previous millennia!

What Galileo found amounted in sum to proof against the Aristotelian picture of the Universe and Ptolemy's model of the Solar System. What Galileo saw proved Copernicus's theory of the Solar System, showing that the Earth was a planet revolving around the Sun, and that the stars lay at immense distances from Earth.

Within a matter of weeks, Galileo had published his observations and his analysis in a book *Sidereus Nuncius* ("The Starry Messenger"). He sent copies to fellow scientists all over Europe, including Kepler, who immediately confirmed Galileo's observations by making telescopic observations of his own, and who wrote a tract in support of him. Others refused even to look through a telescope when given an opportunity – they could not support a book that ruined a cherished and well-integrated theory of life, the universe and everything. If Copernicus's book, *De Revolutionibus Orbium Coelestium*, was about revolutions in the technical, astronomical sense, Galileo's was about an intellectual revolution – something that is never easily accepted.

Above Astronomical Observations: Venus *by Donato Creti (1711). Two astronomers sit and stand on a riverbank at dawn discussing their observations of Venus with a telescope mounted on a quadrant, partly obscured by a tree. A girl in the foreground is resting, shoes off, after a long walk, reflecting thoughtfully on her romantic prospects. The planet itself is a large, thin, featureless crescent.*

9

Cosmographic mysteries

Johannes Kepler (1571–1630) discovered the true shape of the orbits of the planets, using accurate observations made by Tycho Brahe (see page 26). He was born near Stuttgart and intended to pursue a career as a cleric, but in 1594 became instead a mathematics teacher in Graz in Austria. He conceived a model of the Solar System, in which the planets went around the Sun (as Copernicus had proposed 50 years earlier) in orbits whose relative sizes were related to the regular geometric solids (tetrahedron, cube, octahedron, dodecahedron, icosahedron). He published this theory in 1596 as *Mysterium Cosmographicum* ("The Cosmographic Mystery"). Although the connection with the regular solids was spurious, plausible only because of the inaccurate observations that were available at the time, the book showed how Kepler's mind was working, as he sought for the underlying mathematical reasons to explain why the planets moved as they did.

As a Lutheran in Austria at the time the Catholic Counter-Reformation was beginning, Kepler suffered religious persecution when he refused to become a Catholic and in 1600 he moved to Prague to become Tycho Brahe's assistant. Tycho died the following year and Kepler succeeded him as imperial mathematician, where his duties included casting horoscopes for members of the court, including the Emperor Rudolf himself, for whom he added political counsel. He also inherited Tycho's closely guarded high-precision observations of the planets. Put on the right track by Copernicus and Galileo, and working with Tycho's data – particularly his measurements of the planet Mars – Kepler was able to determine the true shape of the orbits of the planets – ellipses, not circles. He was also able to describe the varying rates at which the planets move in their orbits.

With the abdication of Rudolf in 1611, Prague descended into religious strife and Kepler had to leave the city in 1612 for Linz. Even here, he was troubled by religious conflict, forced to defend his mother, who had been accused of being a witch by a malicious woman engaged in a financial dispute with the family. But he was also able to experiment with various laws about the relative sizes of the orbits of the planets to replace his earlier attempt based on geometric solids and found his so-called harmonic law of planetary motion, which he published in 1619, in a book called *Harmonices Mundi* ("Harmony of Worlds").

His three laws of planetary motion, based on the past positions of the planets, enabled Kepler to make more accurate predictions of their positions in the future, which he completed in 1623 as the Rudolfine Tables, named after the Emperor, his former patron.

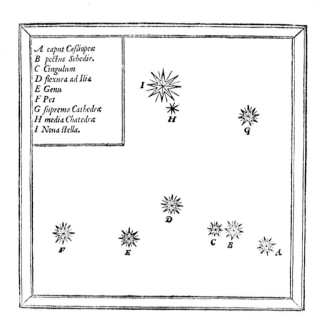

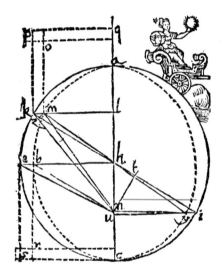

It was a triumph which Kepler did not live to see when the English draper's merchant and amateur astronomer William Crabtree (1610–44) and the cleric Jeremiah Horrocks (1618–41) used these tables to predict the transit of Venus across the Sun on 24 November 1639 (in the Julian calendar). This event would have been totally unexpected using previous calculations but occurred as predicted by Kepler's work.

Left *Kepler's model of the Solar System made up of nested regular solids that just fit into the orbits of the planets.*

Above, top *Tycho's map of the new star that he saw in 1572 marks it as I,* Nova Stella, *with the other stars of Cassiopeia identified in Latin by reference to the constellation figure (A,* caput *(head), F,* pes *(foot), etc).*

Above *Kepler drew the orbit of Mars as an ellipse in his book* Harmonices Mundi.

43

KEPLER'S THREE LAWS OF PLANETARY MOTION

From Tycho's observations of the planets, Kepler showed in his first law that their orbits around the Sun are accurately described as ellipses, not as circles or combinations of circles called epicycles. Furthermore, in his second law, he described how each planet moves more quickly when its elliptical orbit takes it nearer to the Sun and more slowly when it is more distant, in such a way that if you join the Sun to the planet with a straight line, the line sweeps out equal areas of the orbit in equal times. Kepler's third law of planetary motion is that "The square of the orbital period of a planet is proportional to the cube of its mean distance from the Sun".

Kepler's laws describe intriguing mathematical relationships in planetary orbits. These relationships remained unexplained for nearly fifty years until Isaac Newton created his theory of gravity.

Below *William Crabtree viewed the transit of Venus across the Sun in 1639 in the attic of his family's draper's shop. It is a mural by Ford Maddox Brown in the Manchester Town Hall.*

Right *Johannes Kepler in a portrait by an unknown artist (1610). He holds a pair of dividers, used for measuring off globes and maps, a help no doubt for his astronomical calculations.*

Below *Grappling with the consequences of Copernicus' theory that the Sun is one star among many, Johannes Kepler drew the solar system surrounded by a pattern of stars that extend to infinity. He pointed out the problem that arose if this were the case, that wherever you look you should, ultimately, be able to see a star. However, this is not the case and in the night sky there are vast areas devoid of light. This problem has become known as Olbers' Paradox, after the discussion by the nineteenth century German astronomer Heinrich Olbers. It is still troublesome even though we now know that the stars extend only to the edge of our Milky Way, because the same argument goes for galaxies. The reason why the night sky is not ablaze with light is because the universe had a beginning, and we can only see back as far as light has travelled since the universe began. Published in Kepler's* Epitome of Copernican Astronomy *in 1627.*

THE NEW STARS OF 1572 AND 1604

After dinner one evening in 1572, Tycho Brahe was driving home in his carriage. He saw a group of peasants marvelling at something in the constellation Cassiopeia. What he saw where they were pointing was a new bright star in the night sky. Its appearance destroyed Aristotle's belief that the stars were unchanging.

For over a year from his home, then in Heridsvaad, Brahe repeatedly measured the position of the new star (now known as Tycho's supernova) and proved that it did not shift its position in the slightest – its parallax was zero (the parallax of a star is the angle by which it shifts as seen from Earth as this planet rotates or orbits round the Sun). The star had to be at a distance well beyond the orbit of the Moon, in the sphere of the other fixed stars.

The lid was hammered down on the coffin of Aristotle's view that the stars were unchanging by a second nova that appeared 32 years later. It was not Kepler himself who discovered the supernova in the constellation Ophiuchus but in 1606 he collected together the observations that were made of it and it is known by his name.

These two events were the exploding supernovae of 1572 and 1604. In their place today are expanding spheres of the fragments of the stars that exploded, hollow bubbles created in the gas in interstellar space by the explosions.

In 1573 in *De Nova Stella* ("*The New Star*"), Tycho wrote about his discovery:

"Last year in the month of November, on the 11th day of that month, in the evening, after sunset, when according to my habit, I was contemplating the stars in a clear sky. I noticed that a new and unusual star, surpassing the others in brilliancy, was shining almost directly above my head; and since I had, almost from boyhood, known all the stars of the heavens perfectly (there is no great difficulty in attaining that knowledge), it was quite evident to me that there never before had been any star in that place in the sky, even the smallest, to say nothing of a star so conspicuously bright."

Kepler's supernova appeared nearby to a conjunction of Mars and Jupiter, a celestial event that was thought by astrologers to be of great significance, the start of an 800-year cycle. Two previous similar events had occurred at the rise of Charlemagne in about 800 AD and at the birth of Christ, indeed Kepler thought the event might be the origin of the legend of the Star of Bethlehem that had brought the Magi to the scene of the Nativity. The conjunction of 1604 was thus thought to herald something really significant for Emperor Rudolf.

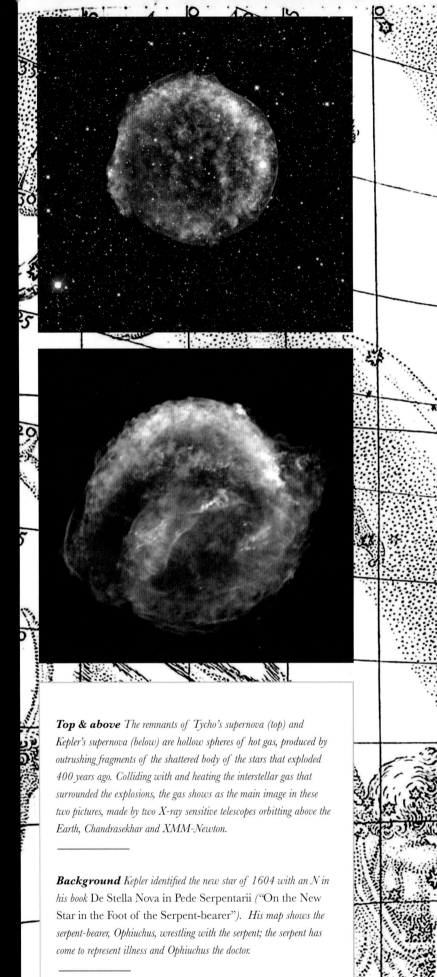

Top & above *The remnants of Tycho's supernova (top) and Kepler's supernova (below) are hollow spheres of hot gas, produced by outrushing fragments of the shattered body of the stars that exploded 400 years ago. Colliding with and heating the interstellar gas that surrounded the explosions, the gas shows as the main image in these two pictures, made by two X-ray sensitive telescopes orbiting above the Earth, Chandrasekhar and XMM-Newton.*

Background *Kepler identified the new star of 1604 with an N in his book* De Stella Nova in Pede Serpentarii *("On the New Star in the Foot of the Serpent-bearer"). His map shows the serpent-bearer, Ophiuchus, wrestling with the serpent; the serpent has come to represent illness and Ophiuchus the doctor.*

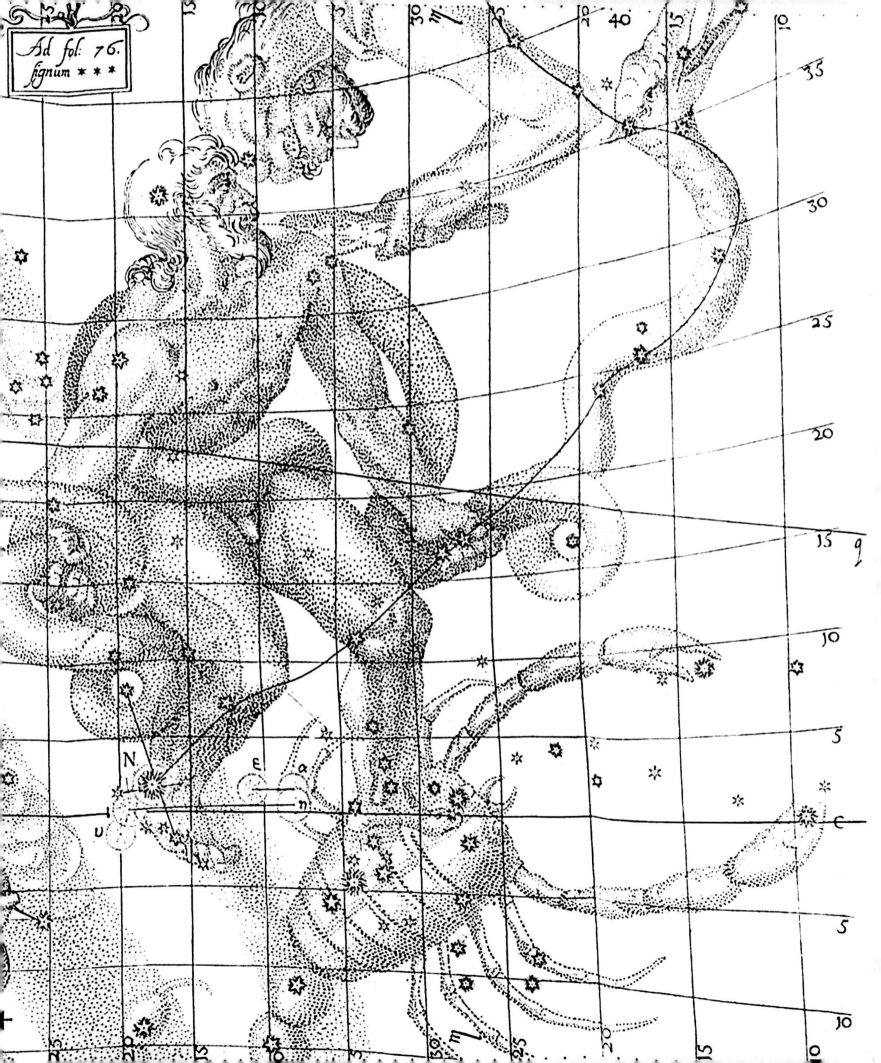

Universal attraction

As Galileo died, Isaac Newton (1643–1727) was born. He studied at Trinity College, Cambridge University. Soon after he was awarded his degree in 1665, the university closed because of the Great Plague. Newton returned home to his family farm, Woolsthorpe Manor, for two years of private study. The story – probably embellished – that Newton enjoyed retelling in his old age was that while he was sitting in an apple orchard, at this time, pondering the motion of the Moon around the Earth, he saw an apple fall to the ground and it suddenly struck him that the same force could be responsible for both. He had discovered the law of gravity, which he formulated more precisely as the force between two bodies is along the line joining them and inversely proportional to the square of their separation.

After a break, Newton began in 1679 to work again on the close relationship between his law of gravity and Kepler's laws of planetary motion. Newton discussed his work with a friend, the astronomer Edmond Halley (1656–1742), who persuaded him to write the work up in an extensive book in Latin, known as the *Principia* (1687). Newton's theory of gravity became the model for scientific laws – mathematically and logically expressed in the *Principia* – like geometrical theorems. Philosophers adopted this ideal as part of the French Enlightenment and Newton became a symbol of rational thought. As such he was reviled by Romantic writers like William Blake (1757–1827) as representing cold, unfeeling Reason.

In a stunning demonstration in 1705 of the power of Newton's theory, Halley determined that the orbit of a comet which had appeared in 1682 was the same as the orbits of two comets of 1531 and 1607. He concluded that all three comets were the same one, returning every 76 years or so, with the next appearance due in 1758. Poignantly this proved to be correct, 16 years after Halley had died.

Right *Newton by William Blake (1804) in a print enhanced by pen and ink, and watercolour. Newton is unsympathetically shown as a perfect god-like figure, coldly measuring a mathematical diagram with dividers. Blake was critical of disciplined reasoning and believed more in the creative imagination.*

In 1703 Newton became President of the Royal Society of London and two years later was knighted, apparently the first person to be so honoured for scientific work. In this influential position he became the chairman of the governing body of the Royal Observatory, which had been founded in Greenwich in 1675 in order to apply astronomical techniques to navigation. Its first director, John Flamsteed (1646–1719), determined the positions of the stars, planets and the Moon to greater accuracy than before, and timed the Sun as it was carried across the meridian to the south of the Observatory by the rotation of the Earth. Flamsteed's scale of time became "Greenwich Mean Time", eventually adopted as the basis for the global system of time zones and of longitude.

Flamsteed started to develop theories to predict the positions of the Moon and planets for navigation, work which eventually turned into an annual volume for sailors called *The Nautical Almanac*. Flamsteed limited his work to the purposes for which the Observatory was founded, and was reluctant to publish imperfect data, but Newton demanded that Flamsteed should immediately communicate data that could establish the scientific theory of gravity. The two men fell out, disputing the priority of the applied or pure science in the work. Halley brokered an agreement that Flamsteed would indeed provide interim data for Newton's private use. One can sympathize with Flamsteed's fury when, walking past a bookshop in London, he found that Newton had published his data in a book.

Right and Below *Edmund Halley and the comet named for him when it reappeared in 1982, its fourth appearance since he predicted that it would return periodically.*

Right The Octagon Room in the Royal Observatory in Greenwich was the original room from which astronomical observations were made, hence the clocks, and high windows for the telescopes to view near to the zenith.

Below The Great Comet of 1532. In a watercolour sketch labelled (top) as the comet of 1532, originally bound in a sixteenth century commonplace book, the comet was shown as lying in front of the clouds in the sky, in accordance with the belief that comets were meteorological phenomena. This particularly bright comet was seen for four months after its discovery at the end of 1532, at times so bright that it could be seen during the day. It was the brilliant comet seen just as the Spaniards arrived in the New World, and superstitiously thought at the time to have heralded the collapse of the Aztec and Inca empires. In Germany, the astronomer Peter Apian observed this comet and noticed that its tail always pointed away from the Sun. There was a suggestion, originated by Edmond Halley, that the comet of 1582 might possibly have been an earlier apparition of one seen in 1661, since their orbits around the Sun were so similar, just as the orbit of the comet which had appeared in 1682 was the same as the orbits of two comets of 1531 and 1607 (Halley's Comet). But it seems that the comet of 1582 was in reality a once-only visitor to our part of the solar system.

PLAN du premier Etage au dessous de la platte forme.

TIME AND LONGITUDE

From the seventeenth century, astronomical observatories were created by powerful trading nations for the purpose of providing a standard of time and of longitude for the mapping of each country. The first was the Paris Observatory, established in France under the direction of the Académie des Sciences in 1667. Coastal maps showed longitude based on meridians through each of these national observatories, but this caused confusion among sailors as they sailed from one map to another. There was a demand for a standardized system of latitude and longitude. Latitude was easy – everyone agreed that it was measured from a zero-point at the Earth's equator. But the zero-point of longitude – the "Prime" Meridian – was not obvious. A second, related concern was the standardization of a system of time zones, to rationalize the systems of time keeping.

A conference was convened by the United States Government in Washington in 1884 to decide the issue of the Prime Meridian. The Paris and Greenwich Meridians were obvious contenders, because of the work that had been put into their definition. However, there was also an argument that a neutral meridian should be chosen, favouring no particular nation – a meridian that ran through some natural or artificial feature (such as the Great Pyramid in Egypt).

There were two decisive practical arguments. Most of the world's cargo was carried in ships flagged to nations of the British Empire, which used the Greenwich Meridian as a standard. Moreover, the railroad companies of the United States operated extensive networks across North America and had a strong need to standardize conflicting timetables. The majority of companies could not agree to base the system on the time kept by the United States Naval Observatory in Washington, since that was thought to give a competitive advantage to companies local to the District of Columbia. Looking at how the various networks connected at important junctions, the companies agreed that the most convenient system of time zones would not be based on Paris. The Washington Conference chose the Greenwich Meridian as the Prime Meridian and Greenwich Mean Time as the basis for the world's time zones.

The modern time-scale is called International Atomic Time (TAI, from the French version of its name). It is no longer based on astronomical observations, which rely on the uniform daily rotation of the Earth – the Earth wobbles too much. Since 1967, time has been based on clocks that count the oscillations of caesium atoms in 200 clocks around the world ("atomic clocks"). But to regulate everyday living, TAI is adjusted by amounts that more or less keep up with the rotation of the Earth – this is Coordinated Universal Time (UTC), the time system that is broadcast by radio; UTC is close to Greenwich Mean Time.

Above *The Royal Observatory in Greenwich, originally built by architect and astronomer Christopher Wren. The building is surmounted by a red globe that drops down the mast visibly to mark the time for ships anchored in the harbour in the River Thames below.*

Opposite *Astronomers manipulate their telescopes in the garden south of the Paris Observatory, and on its roof.*

New eyes on the sky

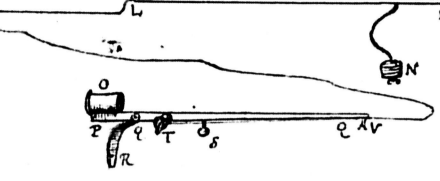

Galileo's telescope was small, its field of view was tiny, its magnification was low and its images were blurry – it is a marvel that it was such an improvement on the unaided eye. But scientists and technologists quickly turned their ingenuity to developing new methods of glass- and lens-making to create larger, clearer lenses.

There was one fundamental problem inherent in lens telescopes ("refractors"). When light is refracted in a lens, it treats each colour slightly differently. Not all colours are focused in the same place. The images are blurred and have coloured edges. This defect is called chromatic aberration.

There are three ways around the problem. The first approach, developed by Christiaan Huygens (1629–95), was to minimize the chromatic effect by having the lenses work less hard – weaker lenses are thinner and not so curved, with less chromatic aberration. Some scientists made extremely long, tubeless telescopes, with the weak front lens slung high on a pole, the image inspected by an eyepiece nearer to ground level. These so-called aerial telescopes were difficult to point and vibrated in the slightest breeze. However, the Italian-French astronomer Gian Domenico Cassini (1625–1712) successfully used an aerial telescope at the Paris Observatory to discover two faint satellites of Saturn.

The second approach to solving the problem of chromatic aberration lay in optical design. Two or more lenses of different sorts of glass could be combined so that the chromatic aberration of one countered the chromatic aberration of the other. Such a combination is called an achromatic lens, invented by an amateur optician, Chester Moore Hall (1703–71) and put into production by another optician, John Dollond (1706–61). The biggest achromatic telescope was the 40-inch (102-cm) telescope built by Alvan Clark (1832–97) in 1895 for the Yerkes Observatory near Chicago.

Right *In 1684 Christiaan Huygens built a tubeless (aerial) telescope, with its eyepiece and light-gathering lens connected by a rope of the right length to focus the image.*

The third and best way to solve the problem of chromatic aberration was to make telescopes from mirrors, for which chromatic aberration does not exist. Isaac Newton was the first to do this, producing a prototype reflecting telescope in 1668. Large mirrors flex too much under their own weight, but engineers developed complicated mechanical designs to support the mirrors at the back surface. The largest, modern, single-mirror reflecting telescopes, with mirrors eight metres (26 feet) across, are the four telescopes of ESO's (European Southern Observatory) Very Large Telescope (VLT) in Chile and the Japanese Subaru telescope in Hawaii. Their mirrors are supported by computer-controlled devices which respond second by second to the shifting strain of gravity. Even larger 10-metre (33-feet) reflecting telescopes (the Keck Telescopes in Hawaii and the Gran Telescopio de Canarias on La Palma in the Canary Islands) have mirrors made of an assembly of smaller mirrors, each aligned with the others by a complicated laser measurement system

Above *The largest refracting (lens-type) telescope in the world, the Yerkes Telescope in Chicago.*

Right *Isaac Newton made a small prototype reflecting (mirror-type) telescope in 1668.*

New planets

William Herschel (1738–1822) was born in Hanover and became a bandsman for the Hanoverian army. At the age of 19, he migrated to Britain and settled in Bath as a musician. His sister Caroline (1750–1848) had taken to heart her father's cruel observation that, poor and scarred by smallpox, she would never marry (in the event a true prediction). In 1772, she escaped domestic drudgery in Hanover, joining William to become his housekeeper and accompanist.

In his spare time, William taught himself astronomy and made a high-performance, functional telescope. He devoted clear nights to a systematic survey of the sky. Caroline acted as his assistant, noting down the nebulae, star clusters and double stars that he found and called out from the eyepiece.

Right Herschel made and sold hundreds of this design of telescope, with a six-inch (15 centimetre) mirror. The telescope has a 7 foot (2 metre) mahogany tube, pulleys, ropes, gears and a stand on castor wheels that can readily be brought out from a house onto a patio. The astronomer stood and looked into an eyepiece at the top of the tube, and used the smaller, wider-angle telescope to sight the larger telescope on the desired object.

Uranus and the asteroids

In 1781, William noticed an object that was not just a point of light. It had a perceptible disc, and moved relative to the background of stars. It proved to be a planet, the first new one discovered since antiquity, lying beyond Saturn. Herschel called the planet Georgium sidus (George's planet) after the then king, George III. William moved closer to the court at Windsor Castle, where he conducted star parties for the amusement of the royal family. His name for the planet did not find favour in countries outside Britain and it became known as Uranus. Meanwhile, Caroline had begun to use her own telescope and in 1786 discovered a comet. Known as The Lady's Comet, it was the first comet discovered by a woman.

Uranus was the first of several new planets discovered over the following centuries. In 1766, German astronomer Johann Titius (1729–96) discovered a formula, popularized by Johann Bode (1747–1826), that connects the radius of a planet's orbit with its number in sequence from the Sun – Mercury is planet number 1, Saturn is number 7. When Uranus was discovered it was found to fit the formula at number 8. But the formula, known as "Bode's Law", only worked if Mars is planet number 4 and Jupiter is number 6. Where was planet number 5?

Right The German-born British astronomer, Sir William Herschel, catalogued double stars, clusters of stars and nebulae during comprehensive surveys of the sky, discovering Uranus as he did so.

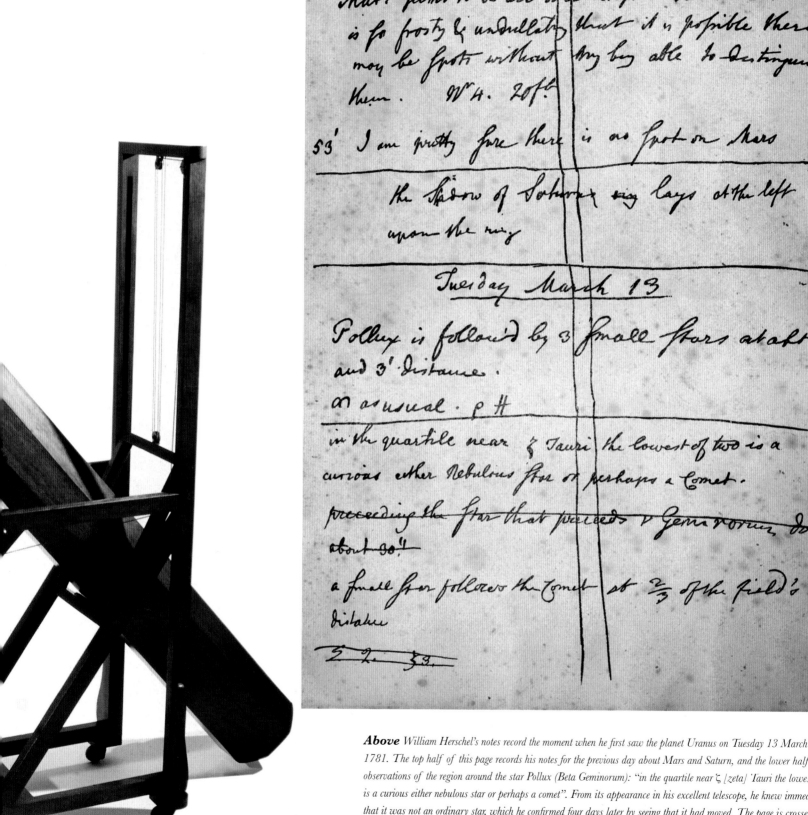

Above William Herschel's notes record the moment when he first saw the planet Uranus on Tuesday 13 March 1781. The top half of this page records his notes for the previous day about Mars and Saturn, and the lower half his observations of the region around the star Pollux (Beta Geminorum): "in the quartile near ζ [zeta] Tauri the lowest of two is a curious either nebulous star or perhaps a comet". From its appearance in his excellent telescope, he knew immediately that it was not an ordinary star, which he confirmed four days later by seeing that it had moved. The page is crossed out by two vertical lines, made by William's sister Caroline to signify to herself that she had transcribed her brother's rough notes into a fair-copy note-book.

The German astronomer Baron Franz Xaver von Zach (1754–1832) organized a group of astronomers to search for the missing planet. What they were looking for turned up on the first day of the new century, 1 January 1801, found, not by a member of the group, but by Sicilian astronomer Giuseppe Piazzi (1746–1826). He called it Ceres, after the Roman goddess of agriculture, patroness of Sicily. It was a surprise when three further small planets (Pallas, Juno and Vesta) were discovered in quick succession, more or less in the same orbit. All four small planets fitted Bode's Law at number 5.

Indeed, so did the hundreds of thousands of minor planets now known, called "asteroids". Some are planets which have been prevented from coalescing into one large planet, stirred up by the force of gravity of the massive planet Jupiter. Many other asteroids are fragments of these bodies, broken up when they collided with one another. From time to time, some of these fragments fall to Earth as meteorites. The larger asteroids are spherical, but according to pictures obtained by visiting spacecraft, the smaller asteroids are irregular, potato-shaped and pitted by small craters caused by meteor impacts.

Opposite *A false-colour view of Uranus by the Hubble Space Telescope revealed its rings and several of its 27 satellites, and brought out not only subtle bands in the methane clouds that encircle the pole of Uranus but also some bright clouds that have floated up into its higher atmospheric layers.*

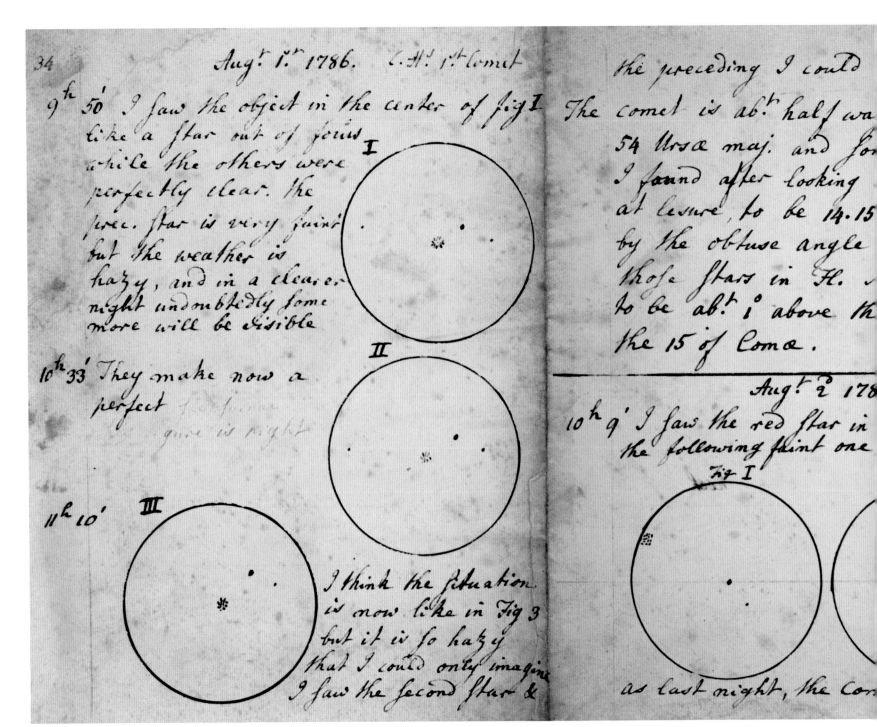

Below left *Caroline Herschel's notes and drawings record the discovery of her first comet, on 1 and 2 August 1786. She noted that the object was "like a star out of focus", frustrated from a clear conclusion by the poor condition of the sky, but later correctly identified it as a comet. The comet and tail are shown in the centre of the three diagrams at left. The comet, now designated Comet C/1786 P1 (Herschel) and familiarly known as The Lady's Comet, was the first of eight comets discovered by Caroline.*

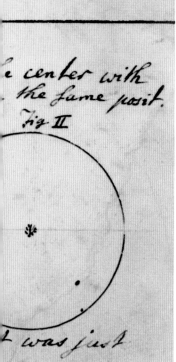

Below *Uranus was imaged in 1986 by the Voyager 2 spacecraft, which showed its almost featureless cloudtops, tinted a light shade of blue by the methane in its atmosphere.*

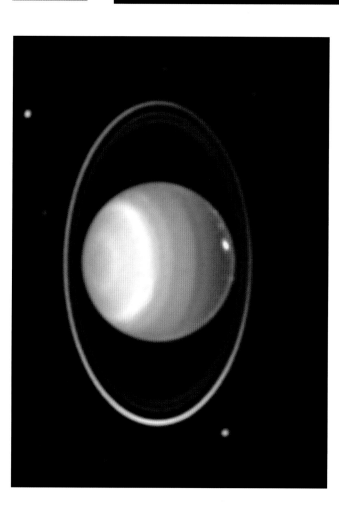

Neptune

Asteroids are minor planets but there was one more major planet to discover, this time located by mathematics. Several historic measurements of Uranus were found, made by astronomers who had mistaken it for a star. This enabled an accurate orbit to be calculated for the new planet. Within decades, Uranus began to deviate from it. In Britain, the Scottish mathematician Mary Somerville (1780–1872) suggested that the reason might be that there was an extra planet outside the orbit of Uranus that was pulling it off track – tugging Uranus ahead when the extra planet was in front, dragging it back when behind. In France, the same thought occurred to the Paris astronomer François Arago (1786–1853). Each of them made the suggestion to a talented young mathematician, on the one hand John Couch Adams (1819–92), on the other Urbain Le Verrier (1811–77), and each of these men independently calculated where the extra planet was likely to be. Adams was socially gauche and failed to inspire any of his professional colleagues to look diligently for the planet, in particular the irascible George Airy (1801–92), then Astronomer Royal at the Royal Observatory, Greenwich. However, Le Verrier communicated his prediction to an enthusiastic young German astronomer Johann Galle (1812–1910) at the Berlin Observatory, who looked immediately, discovering the new planet at the right place on 23 September 1846. The planet became known as Neptune.

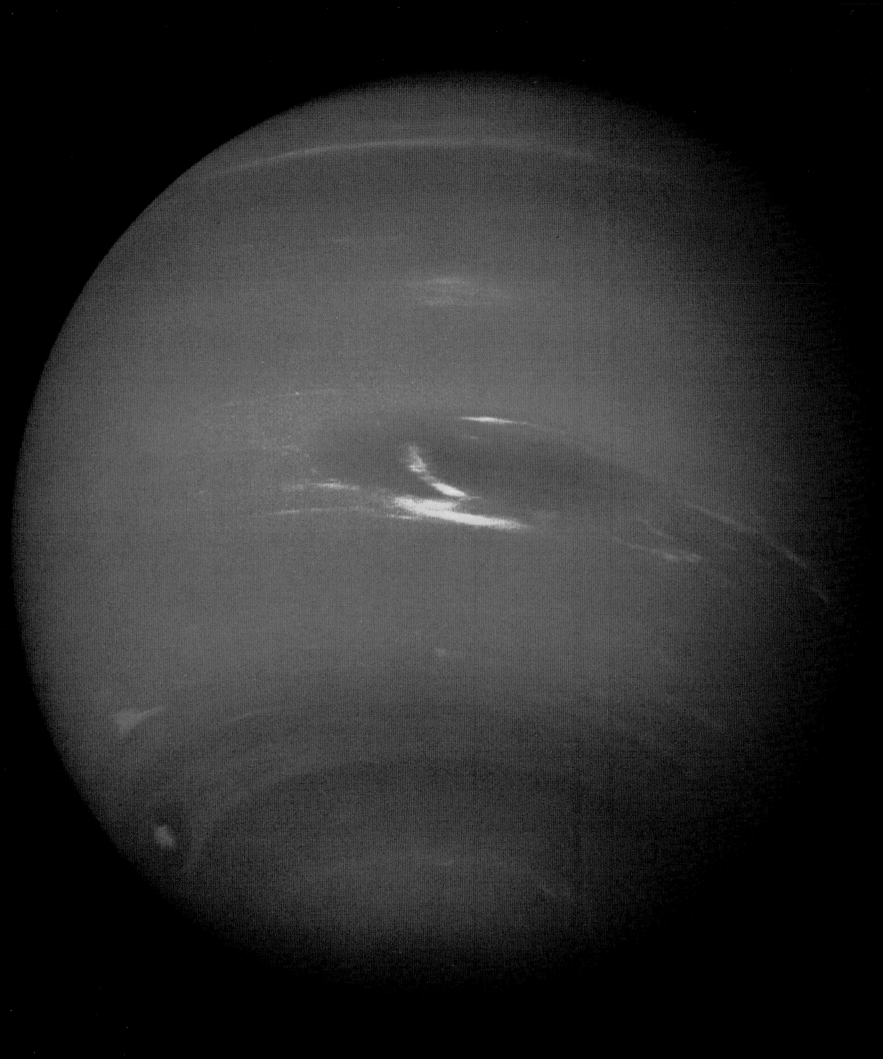

Pluto and the Kuiper Belt

The Solar System had doubled in size with the discovery of Uranus, and approximately doubled again with that of Neptune. Had the edge of the Solar System been reached? As the orbit of Neptune developed, there were reasons to think not – like Uranus before it, Neptune too showed deviations from its predicted path. Based on an analysis of these deviations, in 1906 United States astronomer Percival Lowell (1855–1916) began to search for a ninth planet, "Planet X". In 1929, a young amateur astronomer, Clyde Tombaugh (1906–97), was hired by the Lowell Observatory to continue the search. He photographed portions of the sky on successive occasions a week apart and compared them to see if any "star" had moved. If so, it was not a star at all, but a planet. On 18 February 1930, Tombaugh discovered the new planet orbiting in the cold darkness beyond Neptune. It was named Pluto after the god of the underworld and considered as the ninth planet.

Pluto was much fainter than Planet X had been predicted to be, because it was less massive and therefore reflected less sunlight. In fact, it was too small to have caused the deviations in Neptune's orbit. Pluto had been found only by coincidence at the place where Planet X was supposed to be. Moreover, it was not much beyond Neptune.

In 1943 and 1951, Irish amateur astronomer Kenneth Edgeworth (1880–1972) and United States planetologist

Opposite *Given how featureless the cloud tops of Uranus were, as expected in what was expected to be the cold, weatherless outer reaches of the Solar System, it was a surprise when in 1989 Voyager 2 discovered storms on colder Neptune, including the "Great Dark Spot", an anticyclonic tempest the size of Africa, and other white clouds, that move at speeds up to 2,000 km/hr (1,243 mph).*

Below *Pluto is a world of icy rock, but its surface shows honey-coloured bright and dark regions that are methane ices, broken into tarry goo by the action of ultraviolet light from the Sun. It is only a few pixels across in the Hubble Space Telescope cameras but higher resolution maps were made in 2010 by computation-intensive analysis of a number of images taken over several of its six-day rotations.*

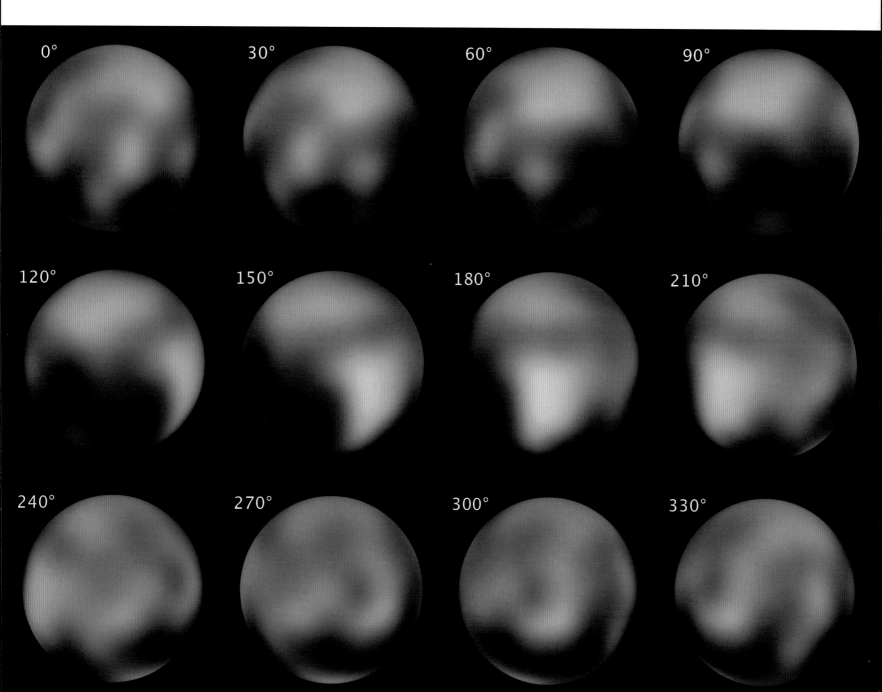

Gerard Kuiper (1905–73), respectively, suggested that there were lots of small bodies in the region beyond Neptune. In 1992, United States astronomer David Jewitt (1958–) and his colleague Jane Luu (1963) found the next so-called Trans-Neptunian Object after Pluto. Now there are over 1,000 known. The region they inhabit is called the Edgeworth-Kuiper Belt. Astronomers looked again at Pluto and decided that its size, orbit and other characteristics were more like a large Trans-Neptunian Object than a small planet. Indeed, one object in the Edgeworth-Kuiper Belt, Eris, is considerably larger than Pluto. For these reasons, since 2006 Pluto has been regarded as a Trans-Neptunian Object rather than as the ninth planet.

Right *Comet McNaught in 2007, one of the brightest comets of the last 50 years, seen here from the Siding Spring Observatory, Australia. Its tail is formed of particles of ice, dust and rocks, exploded in bursts off the comet as it was heated by the Sun.*

Below *The asteroid Lutetia was imaged in 2010 as ESA's (European Space Agency) Rosetta spacecraft flew past at a distance of 3,200 km. It is not much bigger than Death Valley, dry, dusty and pitted with meteor craters.*

COMETS

The region of the Solar System beyond the Edgeworth-Kuiper Belt is the Oort Cloud, the domain of the comets. Extending out to one light year, it really is the edge of the Solar System. It takes its name from Dutch astronomer Jan Oort (1900–1992), who inferred that it existed – it still has not been seen directly.

From prehistoric times, comets have appeared in the night sky, many turning up unexpectedly. They show tails of dust and have traditionally been regarded with superstitious fear as predicting dire events. Halley's Comet was the first comet to be shown to be periodic, after Edmond Halley predicted that it returned at 75 year intervals. It has been seen on every apparition since 240 BC, notably in 1066 when it was associated with the Battle of Hastings.

Comets are masses of ice, left over from the formation of the Solar System. When our Sun passes near to another star or is disturbed by a massive cloud of gas in the Galaxy, some of the icy lumps in the Oort Cloud fall into the Solar System. Warmed by the Sun, the outer layers of ice melt and release dust trapped within; this trails back as the comet's tail. The outer skin of dust fuses into a dark crust as the comet cools on the outward part of its orbit; comets are not the reflective icebergs that one might imagine – they are as black as lumps of coal. If a comet keeps its distance from the planets, it is likely to escape back into the Oort Cloud and may not return to the Sun for millions of years. But if a comet passes near to, say, Jupiter, it may be deflected into a tighter orbit and return periodically.

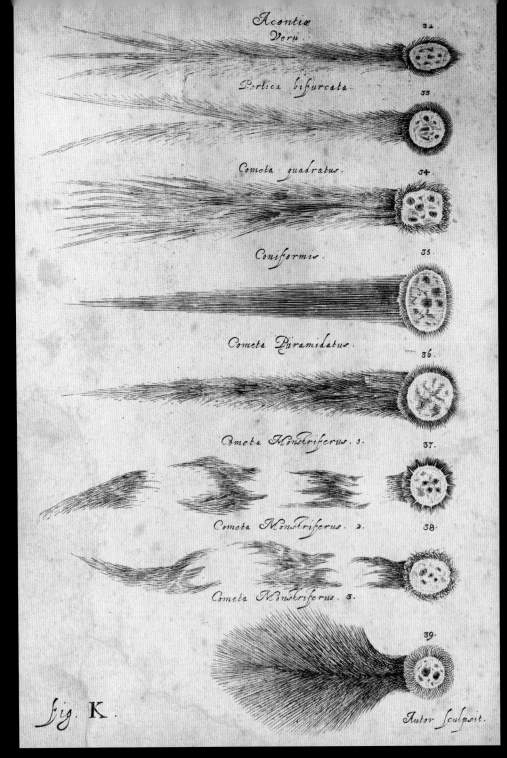

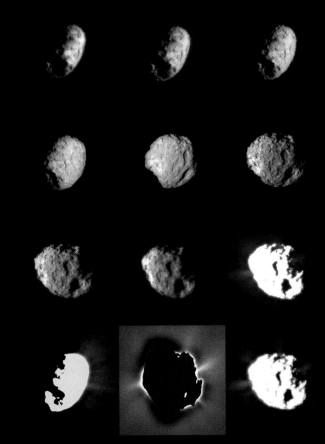

Above An artwork, published in 1668 by the Polish astronomer Johannes Hevelius, shows different types of comets. Comets are bodies of rock and ice that orbit the Sun on elongated orbits lasting hundreds or thousands of years. When a comet nears the Sun, the Sun vaporises the ice in bursts, maybe breaking the comet's head into pieces and creating a long, perhaps irregular tail of dust that aligns on the comet's orbit and reflects the light from the Sun. A comet can also have a so-called ion tail, individual atomic particles that glow and are swept back away from the Sun by solar radiation, along a different path from the dust tail. As a result, comets can have a variety of shapes which Hevelius sketched in his book Cometographia (1668) with Latin labels, some of which refer to bifurcated tails and the variety of cometary heads.

Right Comet Wild 2, as imaged in 2004 by the Stardust space mission as it flew by, in 12 images starting at upper left and continuing left to right and down, like writing. One image is heavily overexposed to show to best effect the jets of vaporized ice that are carrying ice chips, dust and rocks outwards from the comet, feeding its tail. The comet nucleus is shaped like a hamburger patty, seen edge-on in the first images and face-on in the images below.

Above Halley's comet appeared in March 1066, and was depicted in the Bayeux Tapestry, which, strip-cartoon-style, records the events leading up to the conquest of Britain by William of Normandy later that year. Here the English king, Harold, who was killed by Norman archers at the Battle of Hastings months later, is told by a courtier of the worrying appearance of a comet in the sky above his palace. The comet itself is to the right of the words Isti mirant stella[m] (These men are looking at the star).

Right Sir Isaac Newton's own drawing in 1672 of his first reflecting telescope design (1668). It worked by focusing light using a parabolic mirror (V). This reflected light back up the tube of the telescope to another mirror (D) which reflected it into the eyepiece (F). This first telescope was only 15 centimetres (6 inches) long and 2.5 centimetres (1 inch) in diameter but could magnify 30–40 times.

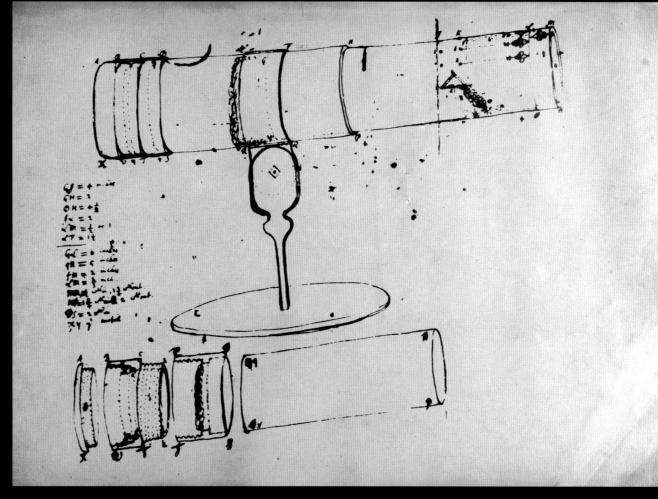

The stars

If the Earth is not stationary but moves around the Sun, why do the stars not move in the opposite direction, just as trees at the track-side appear to move backwards when seen from a train window? The Earth's orbit is negligible in size compared to the distances of the stars, so stars do move, but by small amounts, just as trees a long way away are seen to move less than ones close to the train. This reflective motion of the stars is called "parallax" and measures how far away they are – when our brains use the parallax of things seen by our two eyes we call this "depth perception".

The parallax of stars is very small and throughout history numerous astronomers failed to detect it. The one who first succeeded was Friedrich Bessel (1784–1846), director of the Königsberg Observatory under the patronage of King Frederick William III of Prussia. In 1838, Bessel measured the parallax of the star 61 Cygni, and its distance at about half a million times the distance of the Sun. This knowledge made it possible to calculate the luminosity of the star. This was significant, because it was a property of the star itself, as opposed to its position – a function of astrophysics, as opposed to astronomy.

Another stellar property investigated by astrophysicists was colour. Some stars look red to the naked eye, like the star Betelgeuse in the constellation Orion, some look white or blue, like its neighbour Rigel. There are variations in the balance of light from star to star; red stars emit more red light than blue, blue stars more blue than red.

Astrophysicists were aware that an object like a metal poker would glow red hot, orange, yellow and white hot as it was heated to a higher temperature. The stars must be of different temperatures, too. Given the temperature and the luminosity of a star it became possible to calculate its size. Some stars proved to be much larger than the Sun, even large enough to engulf the entire Solar System.

The astrophysicists who participated in these studies came from a range of backgrounds that were typical of many of the people who at that time studied astronomy: Joseph von Fraunhofer (1787–1826) was an optician, inventor of the spectroscope, and director of the Optical Institute at Benediktbeuern, a former Benedictine monastery; Father Angelo Secchi (1818–78) was a priest of the Jesuit university in Rome; William Huggins (1824–1910) was an independently wealthy British amateur astronomer; Edward Pickering (1846–1919) was a Harvard university teacher; Hermann Vogel (1841–1907) was the director of the state-financed Potsdam Observatory, the first observatory to be devoted to astrophysics.

These scientists improved their techniques and found that the spectrum of light from the stars was overlaid with a multitude of small gaps, called "lines". This phenomenon was interpreted by German physicists Gustav Kirchhoff (1824–87) and Robert Bunsen (1811–99). The hot surface of a star is covered by a cooler atmosphere and the lines are the signatures of the atoms in the cooler gases that absorb particular colours of light. This makes it possible to find out what stars are made of. At last there was real proof that the the stars and the Earth are made of the same materials. As eventually proved by the pioneer astrophysicist, Harvard-based Cecilia Payne-Gaposchkin (1900–79), stars are principally composed of hydrogen, with smaller quantities of all the usual atoms found on Earth.

Above *Father Angelo Secchi, Jesuit priest and astronomer, a founder of stellar spectroscopy.*

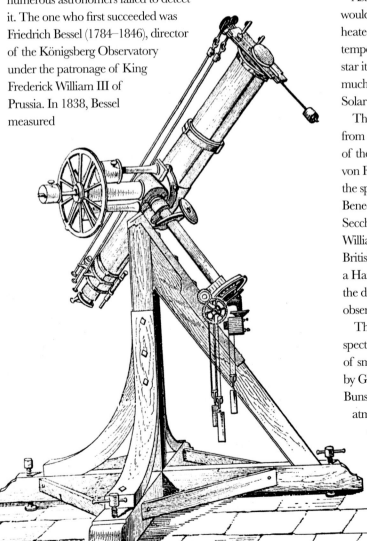

Left *Friedrich Bessel made the first measurement of the distance of a star using this "heliometer", an instrument used primarily to measure the diameter of the Sun as it changed size during the annual motion of the Earth around its eccentric orbit. Bessel put it to use at night time to measure the shift of the star 61 Cygni relative to others as the Earth moved across the diameter of its orbit.*

Opposite *The constellation of Orion has two bright stars of highly contrasting colour. Betelgeuse is upper left and is one of the reddest stars whose colour can be discerned with the unaided eye. It contrasts with Rigel, lower right, which is diamond blue-white.*

On detailed examination of the position of the lines in the spectra of some stars, Pickering and Vogel made a dramatic discovery in 1887. The lines shifted back and forth periodically. It turned out that this was caused by the motion of two stars that were in orbit around each other. These stars were like the more separated pairs of stars discovered by William Herschel, a century earlier. It became possible to determine the mass of the components, using Newton's theory of gravity. Some of the stars were much more massive than the Sun, up to 100 or even 300 times greater. Many were considerably lighter, as small as one-tenth the mass of the Sun.

Gradually, astrophysicists compiled intrinsic data of this sort about the stars. Trying to make sense of it, the Danish astronomer Ejnar Hertzsprung (1873–1967) and Princeton astrophysicist Henry Norris Russell (1877–1957) plotted a graph, star by star, of luminosity against temperature. The graph proved so powerful that it became known as the Hertzsprung-Russell Diagram. It

Above left *Hermann Vogel, German chemist and spectroscopist and director of the above observatory.*

Left *Edward Pickering, American physicist and spectroscopist and Harvard teacher.*

Above *The Astrophysical Observatory of Potsdam, the first founded for astrophysical research.*

clarified the structures of stars: most of them ranged systematically from bright, high-temperature, massive, blue stars to faint, low-temperature, lightweight, red stars. This systematic trend was called the Main Sequence and the stars were termed "dwarfs".

A number of stars fell off the Main Sequence – bright, red, cool stars, called "giants" or even supergiants". Russell found one member of another strange class of star. It was hot and very small, the mass of the Sun packed into a sphere the size of the Earth – a "white dwarf".

It was British a astrophysicist who explained all this: Arthur Stanley Eddington (1882–1944) who explained all this, by relating what could be seen of the outside of a star to its internal constitution. A star supports itself against its own downward force of gravity, by pushing up with the internal pressure of the gas of which it is made. What Eddington's calculations revealed was that the core of a star was very dense and hot, with the temperature and density of the body of the star trailing off to low values at the surface. The conditions in the core were extreme – very hot, very dense. Of course, this raised the question of how these conditions were maintained.

The conditions in the outer layers of some stars are not so extreme, but produce a fascinating effect: some stars throb like a heartbeat. The first example of such a star, Omicron Ceti, was considered so extraordinary that it became known as Mira, the Wonderful. In 1596, the German astronomer and theologian, David Fabricius (1564–1617), noticed that after increasing its brightness considerably, the star faded out of sight. It comes and goes pretty regularly, with a period just short of a year. Mira is an example of a pulsating variable star, which swells and contracts. Its change of size and temperature makes it change brightness. The gas in the outer layers of Mira forms a kind of valve that is opaque when the star is compressed, so that the internal pressure builds up and the star expands. The gas then becomes transparent and lets the radiation out, so the star contracts, for the cycle to start all over again.

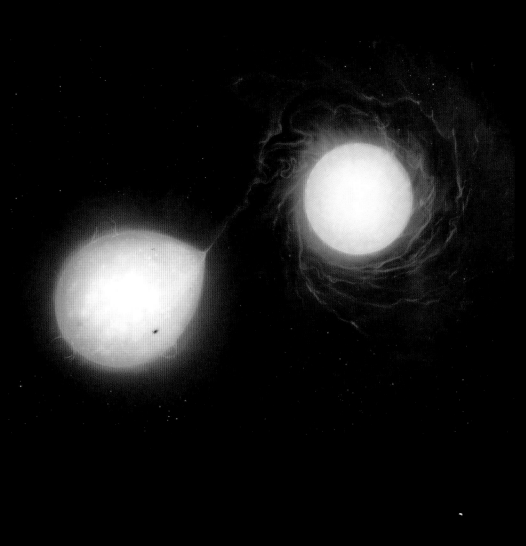

ECLIPSING STARS

In a few cases, the orbit of a double star is so exactly edge-on to the Earth that one star passes across and obscures the other – a form of eclipse. The combined light from the two stars suffers a repetitive, periodic dimming. The first known example of such a star was Algol, the star Beta Persei. Its Arab name, which means "the Demon", and its position in the Gorgon's head of the constellation Perseus, implies that something peculiar about it has been known for some time, but its variability was first recorded in 1667 by the Italian astronomer Geminiano Montanari (1633–87). The star was observed diligently by John Goodricke (1764–86). Born in Groningen, in the Netherlands, to an Anglo-Dutch couple (a British diplomat and a merchant's daughter), he contracted scarlet fever at the age of five and lost his hearing, but overcame this handicap and turned to science. Viewing Algol from a window in his house, he determined that it dimmed every 68 hours and 50 minutes. In 1783, he suggested that this could be caused by a dark body passing in front of the star, as proved to be the case.

Left *Artwork showing the double star, Algol. When the orange star passes in its orbit in front of the smaller, blue one, the combined light from the pair dims considerably. This was how Goodricke discovered that the single star that could be seen by eye was in fact two, one periodically obscuring the other.*

FAINT MEANS FAR MEANS LONG AGO

Light takes eight minutes to travel from the Sun to the Earth. If 61 Cygni is half a million times further, its light takes 11 years to reach us. The furthest stars in our own Galaxy of stars are so distant that their light takes hundreds of thousands of years to reach us. Beyond the confines of our Galaxy lie other galaxies: the light from the nearest of these takes millions of years to get here. The most distant galaxies are very faint indeed, so faint that the Hubble Space Telescope took more than a week to expose their image, in what have become known as the Hubble Deep Field (1995) and the Hubble Ultra-Deep Field (2003–4). These galaxies are 1,000 million million million times fainter than the Sun and light from them has taken nearly the age of the Universe to reach us (about 13,000 million years).

Left *Virtually all the images in the Hubble Deep Field are galaxies, some relatively large, bright and nearby, many small, faint and far away. The fainter galaxies tend to be the red ones, their light redshifted (see page 63) by the expansion of the universe.*

Opposite *The constellation Perseus. The star Algol is just below centre. Also known as Beta Persei, it is the second brightest star in the constellation, and easily visible to the naked eye.*

14

The life of stars

We feel the heat from the Sun and we see its light. Its energy cannot last forever, so what is the Sun's age? The first attempts at an answer were based on the age of the Earth – since the Sun and Earth are so closely connected, the age of the one gives clues as to that of the other. The first approaches assumed that the Earth was hot when made, from material drawn out from the Sun. Yet this led on to the question of how long had it taken to cool? In 1779, the French naturalist Georges-Louis Leclerc (Comte de Buffon, 1707–88) experimented with a small-scale model of the Earth made from a small, hot, iron sphere, and in the second half of the nineteenth century, the British physicist William Thomson (Lord Kelvin, 1824–1907) made theoretical calculations. Their estimates – at 75,000 years and 40 million years respectively – were gross underestimates of the true age of the Earth.

German astrophysicist Hermann von Helmholtz (1821–94) and the Canadian-American astronomer Simon Newcomb (1835–1909) took a direct astrophysical tack by assuming that the source of energy for the Sun was the heat that it liberated as it

settled under gravity. How long would the Sun take to contract to its present size, starting from the nebula of gas from which it was born? The answer, 20 million years, was similar.

The modern technique to measure the age of the Earth originated in the first years of the twentieth century from work in Canada by the nuclear scientists Ernest Rutherford (1871–1937) and Frederick Soddy (1877–1956) on radioactive elements that break down over time into other, daughter elements. Radioactive radium and its daughter, helium, are found in the Earth's crust and Rutherford calculated how long it had taken for the one to decay to the other: it was 40 million years. Applying new, detailed knowledge about radioactive decay gathered in the first half of the twentieth century, geologist Arthur Holmes (1890–1965) concluded from old rocks from all over the world, including Sri Lanka, that the Earth was 1.6 billion years old. The so-called Genesis rock retrieved by the Apollo 15 astronauts from the surface of the Moon is 4.5 billion years old. The most modern figure for the age of Earth is 4.6 billion years, and that the Sun is a little older, being the Earth's parent body.

The amount of energy released over the lifetime of the Sun is prodigious. What is its source? The new, twentieth-century science of nuclear physics posed the problem and also provided the answer – the fusion of four hydrogen nuclei to one helium nucleus, a nuclear process that scientists have mimicked in the laboratory and which they seek to carry out on a large scale as a terrestrial energy source.

There is a lot of hydrogen in the Sun, which for billions of years has provided the light and heat energy that we see and feel. But stars, though big, are of course finite, and eventually the hydrogen in the core of each star runs out. What happens next was discovered by a number of astrophysicists using computer calculations developed in the second half of the twentieth century from nuclear bomb codes. The centre of the star's core compresses further and the core itself grows larger. This triggers a further nuclear reaction, in which three helium nuclei – which had resulted earlier from hydrogen fusion – themselves fuse to make a carbon nucleus.

Above left *The Genesis Rock was brought to Houston and became sample #15415.*

Below left *Ernest Rutherford, physicist and Nobel Laureate.*

Opposite *Apollo 15 landed in Mare Imbrium at the base of the Apennine Mountains. In their second geological exploration of the Moon's surface, on 1 August 1971, astronauts Dave Scott (the Mission Commander) and Jim Irwin (the Lunar Module Pilot) were foraging in moon-dust near the rim of Spur Crater when Scott spotted a rock that took his fancy: "Okay. Now let's go down and get that unusual one. Look at the little crater here, and the one that's facing us. There is a little white corner to the thing." Irwin photographed it in situ, with the calibration device called the gnomon standing over it. Irwin photographed his own shadow and Scott's legs as he waited near by. Scott picked the rock up with long-handled tongs and they looked at it, shaking the dust off. Scott laughed as it glinted. "Oh, boy! I think we might have ourselves something crystalline. What a beaut!" They put what became known as the Genesis Rock in a sample bag, to carry it with them back to Earth.*

As a result of these internal changes, the star has to adjust its structure. It expands and becomes a giant star and may go on to be a supergiant. Further changes take place as the helium runs out and the star uses the carbon as a fuel to make oxygen and sulphur. Thus stars change from one sort of star into another, progressively throughout their lives; the changes taking place relatively quickly after longer periods of stability. Our Sun is in the middle of its first long stable stage, something which has enabled life to evolve on Earth over the past 4.6 billion years, but after another five billion years, it will swell to become a red giant, engulfing the Earth and the other planets.

Right *A tokomak device at the Joint European Torus, which uses magnetic fields to confine hot, hydrogen plasma that would melt any solid container. The hydrogen is stimulated so that its nuclei fuse and release fusion power.*

Overleaf *Cassiopeia A is a supernova remnant from a star that exploded in about 1680, unnoticed by any astronomers except for one accidental observation of a star that is not now there. The elements that were the products of nuclear fusion in a star have been scattered into space when it exploded and show as different colours in the nebula that resulted – blue light comes from oxygen, red from sulphur.*

NUCLEAR ENERGY IN THE SUN

Two atomic physicists of the University of Göttingen, Fritz Houtermans (1903–66) and Robert d'Escourt Atkinson (1898–1982) worked out the source of the Sun's energy on a walking tour during their summer holiday in 1927. Eddington had calculated the density and temperature inside the Sun. In these conditions, the atoms in the centre of the Sun frequently collided together, so hard that they became broken or "ionized" – even their nuclei banged together very hard. This provided the opportunity for nuclear processes to change the nuclei from one kind to another and release nuclear energy. Atkinson later learned that the Sun was mainly hydrogen and realized that the source of the energy was nuclear fusion of hydrogen. German physicists Hans Bethe (1906–2005) and Carl von Weizsäcker (1912–2007) filled in further details in 1939, work for which (with other discoveries) Bethe was awarded the Nobel Prize in 1967.

MASS AND ENERGY

A helium nucleus is 0.7 per cent lighter than four hydrogen nuclei, so if four hydrogen nuclei are fused to a helium nucleus, this mass converts to energy according to Einstein's famous formula, which he formulated in 1905, $E = mc^2$. The fusion of just four hydrogen atoms provides a minuscule amount of energy, but multiplied many times over by the astonishing number of hydrogen atoms in the Sun, this process provides the output of the Sun: 400 million tonnes of mass disappears from the Sun every second and is transformed into solar energy.

Left The first hydrogen bomb, known as Ivy Mike, detonated on 1 November, 1952 on Enewetak, an atoll in the Pacific Ocean, releasing enormous amounts of fusion energy.

The death of stars

Stars end their lives in different ways but always from the same cause: starvation. As its nuclear fuel runs out, the star has reduced "food" and it dies. Without heat, the star can no longer prop itself up against its own force of gravity and something happens to end its life. Quite what depends on the individual star – the key parameter is its mass.

Most stars, those like the Sun, or a bit more massive, expand to become red giants. Because such stars are big, the force of gravity on their surface is weak, and even moderate storms on the surface – which show up as the spots on the surface of the Sun – or in an extreme case, like those on the red supergiant Betelgeuse, cause the star to lose some of its mass into space. If this happens consistently enough, the hot, dying core of the star is exposed. The core radiates extreme ultraviolet light, which lights up the lost material. The phenomenon is called a planetary nebula.

Planetary nebulae were discovered by William Herschel *(see page 44)* while he was making his survey of the heavens. The first one that he stumbled across in 1782 reminded him of his earliest view of the planet Uranus, which is why he called it a "planetary nebula". Such nebulae really have nothing to do with planets, it is just that some of them – such as the Ring Nebula – resemble one. Others have bizarre shapes which have inspired fanciful names: the Butterfly Nebula; the Owl; the Eskimo; the Cat's Eye.

In 1790, Herschel came across NGC 1514, a planetary nebula with a central star. Such a star cools to become a "white dwarf", the kind of small star discovered by Russell in the Hertzsprung-Russell Diagram *(see page 52)*. When first formed, white dwarfs are hot enough to radiate ultraviolet and X-rays, so they are easily picked up in X-ray telescopes. Eventually, white dwarfs cool and disappear from view altogether.

The nature of white dwarfs was explained by two young physicists: Subrahmanyan Chandrasekhar (1910–95) built on work by the Ralph Fowler (1889–1944). The findings startled astronomers, because they revealed that white dwarf stars were supported by a new kind of pressure, called "electron degeneracy", which had been postulated by the then new science of quantum mechanics. Quantum mechanics had been created

Far right *Many stars are accidentally sprinkled all over this picture, but one is centrally on the axis of the "bipolar" structure of the planetary nebula NGC 2346. It is in fact a double star, one of which has created two lobes looking like butterfly wings. In three dimensions they are shaped like two glass tumblers held base to base.*

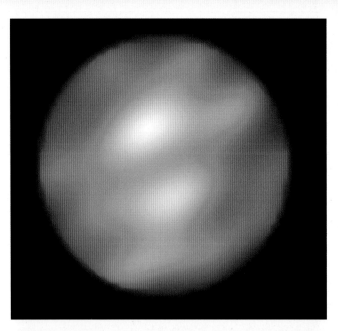

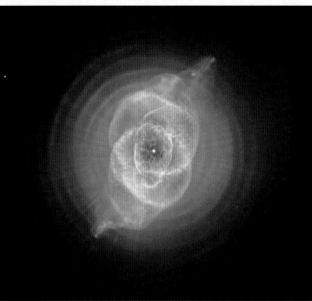

Top *Most stars are just points of light but Betelgeuse is a large enough, close enough star that astronomers can discern some of its surface features – dark spots and bright flares. It is a red supergiant star, past its mature prime.*

Above *Central to the Cat's Eye planetary nebula is its progenitor star, formerly a star much like the Sun and then a red giant, during which stage it puffed off successive spheres of material, one every 1,500 years. About 1,000 years ago the red giant made the transition to the hot white star that it is now, it created a highly structured nebula, with two contorted transparent lobes that overlap as seen from the Earth but which point in opposite directions.*

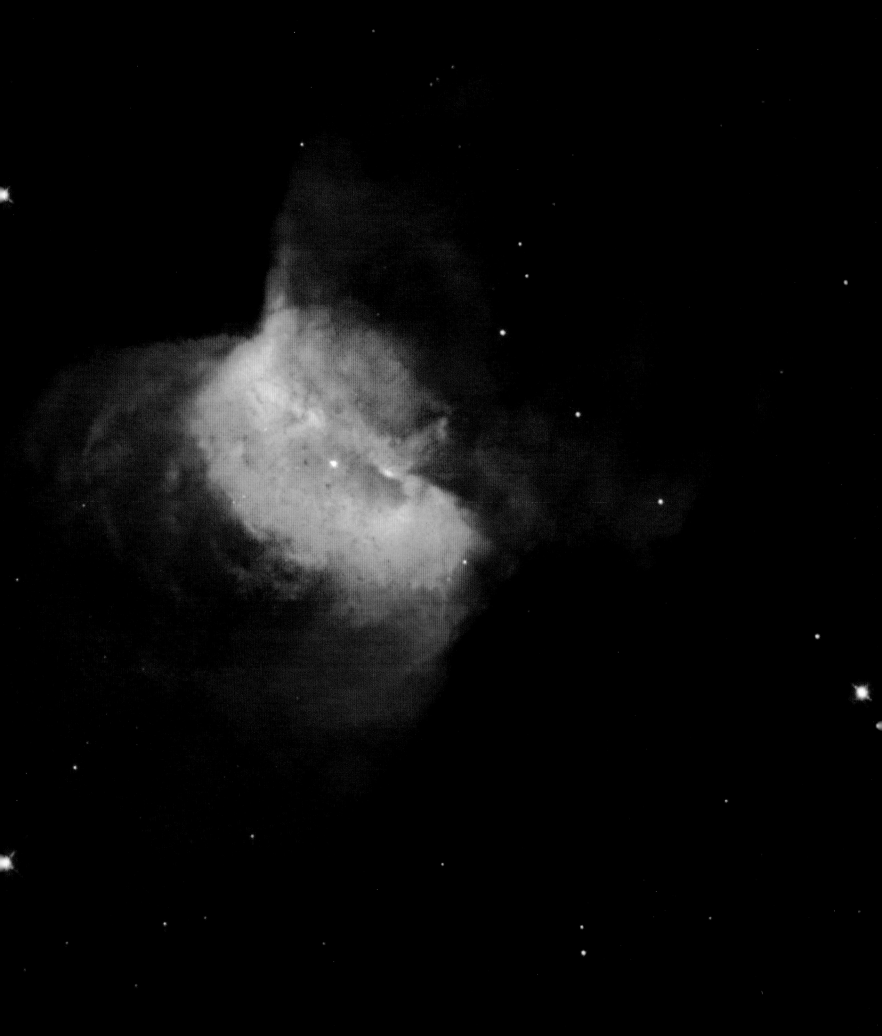

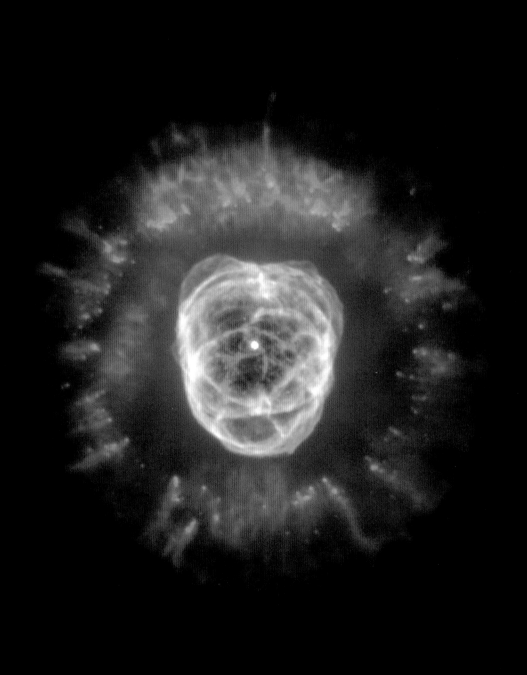

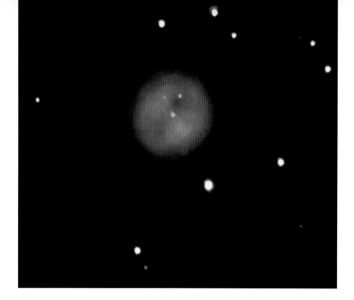

to solve problems in the physics of tiny atoms, and it was a surprise to find that its subtleties were needed to explain some of the biggest objects – stars.

In 1930, Chandrasekhar applied the theory of electron degeneracy to white dwarfs in order to pass the time on a long sea voyage from India to Britain (these were the days before mass air-transportation). He found a surprising relationship between the mass of a white dwarf and its size – as the mass increased, its size diminished. Over a certain mass (called the Chandrasekhar Limit, about 1.4 times the mass of the Sun), white dwarfs are infinitesimally small. If nature starts to make a white dwarf that would be over this limit, it collapses into something with infinite density. This impossibility is called a singularity, and foreshadowed the recognition of black holes.

In Britain, Chandrasekhar met great resistance to his theory, because it contained this impossible outcome. One critic was the senior astrophysicist, Arthur Stanley Eddington. The fierce attack caused Chandrasekhar, an unassuming, young man, a crisis of self doubt. Seeking to stabilize his career, he migrated to the United States and worked for the rest of his life at the University of Chicago, vindicated not only by seeing his work eventually understood by scientists, but also by the award of the Nobel Prize in 1983. But the theoretical problem of the singularity of black holes remains unsolved to the present day. Something must happen inside a black hole as a too-massive white dwarf approaches the infinitely small, but no one knows what it is.

Opposite *Eskimo Nebula. Before the Hubble Space Telescope produced this more detailed picture, the brighter part at the top of this nebula looked like the fur fringe of an Eskimo's parka hood, hence the name. Within the spherical structure there lies a small planetary nebula created relatively recently. The central star was also created recently and is still hot, not having cooled much, so it is pumping out intense ultraviolet radiation, which streams the lumps in the earlier shell into "comets".*

Above *Owl Nebula. In a small telescope, this planetary nebula is a pale, circular disc and looks quite "planetary", but in larger telescopes two dark shadows look like eyes and give it its name. It is probably spherical overall, but the two dark eyes are hollows within the sphere, so there is a bipolar structure embedded within.*

SMALL AND HEAVY

In the theory of general relativity, when light (or other radiation) moves up in a gravitational field it loses energy. This effect, predicted by Einstein in 1911, is called a "gravitational redshift", because blue light shifts in colour towards the red. The effect was verified on Earth in an exquisitely accurate laboratory-scale experiment in 1959–65 by Robert Pound (1919–2010) and his student Glen Rebka (1931–) at Harvard University. The surface gravity of a white dwarf star is much larger than that of the Earth, so the effects of the star's surface gravity are larger and more obvious than the effects of the Earth's, which makes up for the difficulty of investigating something to do with a faint star compared to investigating something in a laboratory.

The first attempt to measure the gravitational redshift of a white dwarf was made in 1925 on the star Sirius B by Walter S. Adams (1876–1956) at Mount Wilson Observatory, California. Sirius B orbits the much brighter star Sirius A, and from 1930 to 1950 the two stars were so close together that the light from the white dwarf was totally swamped. Adams's tentative result was confirmed in 1971 by Jesse L. Greenstein (1909–2002) with the 200-inch (508 centimetre) telescope on Palomar Mountain. In 2004, on the next orbit, astronomer Martin Barstow had the Hubble Space Telescope obtain sharp images of Sirius B, free from contaminating light. The phenomenon of the gravitational redshift was proved. Sirius B was 0.978 times the mass of the Sun and 0.0086 times its radius, or about the radius of the Earth.

Above *The Sirius star system, seen in the Chandra X-ray telescope. The white dwarf star, Sirius B, is hot and emits copious X-rays, so, unlike the system as seen in optical telescopes, it is the brighter of the two stars (with a spike-like structure due to internal struts that support the optics of the telescope).*

16

New windows on the universe

In the 1930s the Bell Telephone Laboratories, physicist Karl Jansky (1905–1950) was investigating radio "static" as a limitation on short-wave radio telecommunications. He built an antenna mounted on a turntable; it was nicknamed the "Merry-Go-Round". Jansky identified a faint hiss, whose intensity rose and fell once a day: could it be the Sun? In fact, its period was not 24 hours – the time of the Earth's rotation relative to the Sun – but 23 hours and 56 minutes, the period of Earth's rotation respective to the stars. The noise came from the constellation Sagittarius, the centre of the Milky Way.

Apart from pioneering work by the US amateur astronomer Grote Reber (1911–2002), the newly founded science of radio astronomy lay dormant for a decade, until the Second World War. Then, during the development of radar in Britain, further sources of celestial radio noise were noticed by defence scientists such as James Hey (1909–2000): active sunspots on the surface of the Sun, and meteors. Once the Second World War was over, some radar scientists initiated "radio astronomy", the first field of astronomy to use a spectrum other than light as its window of study.

Radio astronomy revealed new phenomena. This encouraged astronomers to work with engineers to open other spectral windows, more or less opaque to celestial radiations, like the infra-red and X-rays. X-ray sensitive telescopes, made by United States Nobel Prize-winner Riccardo Giacconi (1931–), were launched around 1960 on small rockets above the Earth's atmosphere to study the Sun. In 1962, Giacconi's team launched an X-ray telescope to observe the Moon and instead found the first source of X-rays outside the Solar System, in the constellation Scorpius. They found another source in Cygnus; the two sources of X-rays became known as Scorpius X-1 and Cygnus X-1, the first of many thousands now known. It was hard to study celestial sources in the few brief minutes of a rocket flight, so X-ray astronomers went on to build satellites on which to fly their telescopes – UHURU (1970), Einstein (1978), Chandra (1999) and XMM-Newton (1999) were major examples among dozens.

Right *For eight hours, astronaut Michael Good worked in one of three spacewalks to refurbish and upgrade the Hubble Space Telescope during its last Servicing Mission in May 2009. Here he is making adjustments while he stands in a foot restraint on the end of Space Shuttle Atlantis's remote manipulator, silhouetted against the Earth below. Hatches on the HST stand open for him to access its inner workings. Astronaut Mike Massimino (lower right, partially out of frame) is busy in the cargo bay.*

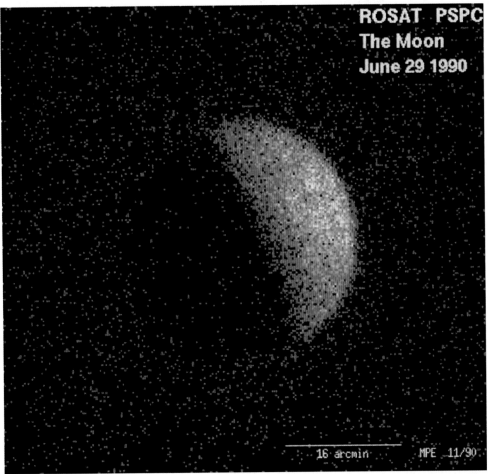

Above *Karl Jansky stands before his rotatable aerial called the Merry-Go-Round. In the 1930's scientists dressed much more informally than the dress-code of 50 years before dictated (see pages 52 and 56).*

Left *The first celestial images made with rocket-borne X-ray sensitive telescopes of the Moon failed to detect anything, but the more sensitive X-ray telescope on the satellite ROSAT saw abundant X-rays from its sunlit side. The X-rays are produced by solar radiation irradiating the Moon's surface, unprotected by any atmosphere.*

HUBBLE SPACE TELESCOPE

Starlight has to pass through air to reach a telescope on the ground. Air disturbs the passage of light and celestial images are naturally blurry. This effect can be mitigated – but not eliminated – by observing from high mountains. In 1990, NASA launched the Hubble Space Telescope, putting a large optical telescope outside the atmosphere for the first time. At first only a qualified success – because it had been made with faulty optics – HST was planned to be serviceable in flight and the fault was corrected in 1993 in the first of five service missions (the last taking place in 2009). The HST will thus continue to be an outstanding discovery machine for about 25 years. In about 2015 it will be replaced with the James Webb Space Telescope, sensitive to infrared radiation.

Below *The Hubble Space Telescope in space as viewed from the Space Shuttle during a service mission.*

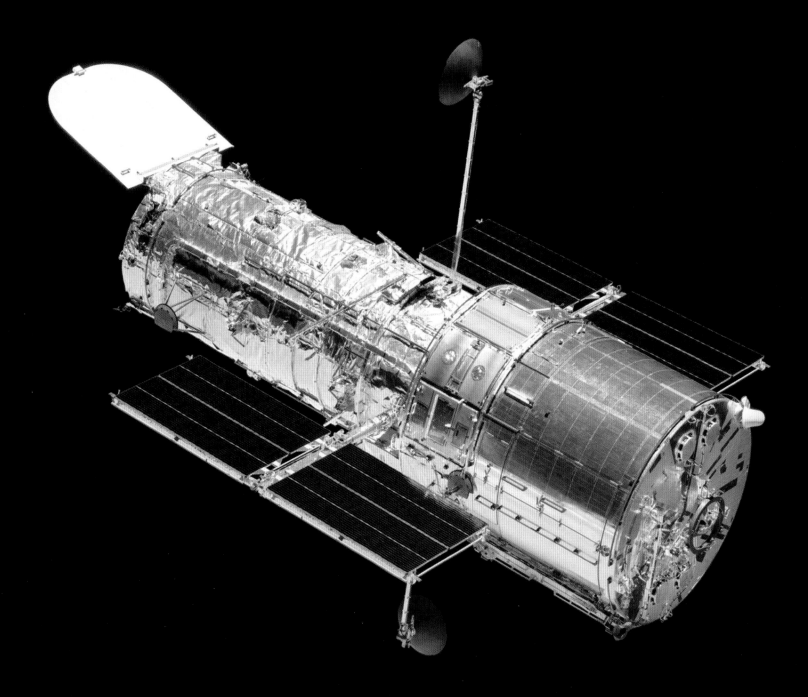

Exploding stars

Most stars end their lives as white dwarfs, but not the larger ones that are over about five or six times the mass of the Sun. If a larger star tries to create a white dwarf that is too massive it cannot – the putative white dwarf star implodes. The energy that the collapse releases blows off the outer layers of the star in a fierce explosion, called a "supernova". A Swiss-American astronomer named Fritz Zwicky (1898–1974) and his German-American colleague Walter Baade (1893–1960) coined the term in 1934 to explain the highly energetic outbursts of light that they had observed in other galaxies. They saw bright stars appear suddenly and without warning, outshining (for a few days) the light from all the rest of the billions of the stars in the galaxy put together. The stars faded away as the explosions dissipated.

Supernovae occur once every few decades in a galaxy like ours. The most famous is the supernova of 1054. It produced not only an outrushing nebula, called the Crab Nebula, but also an energetic star, the stellar remnant of the supernova. It was discovered in 1968 by David Staelin and Edward Reifenstein at the National Radio Astronomy Observatory in Green Bank, West Virginia. It is a pulsar.

"Pulsar" is an acronym for "pulsating radio star" – one which emits pulses of radio waves. The first of these stars was identified in 1967 by Jocelyn Bell (1943–) during her Ph.D. project with a radio telescope in Cambridge supervised by Antony Hewish. The astonishingly regular radio pulses seemed artificial, so, although they were manifestly in interstellar space, the radio-astronomers with whom Jocelyn Bell worked half-seriously contemplated whether they were communication devices like navigation beacons made by extraterrestrial intelligence. They proved in fact to be tiny, rotating stars, sometimes spinning in less than a second, which beamed out radio waves that swept across the Earth on each rotation. The stars are somewhat similar to white dwarfs, but they are even smaller – perhaps 12 kilometres (7.5 miles) in radius, and composed of neutrons.

Yet even neutron stars are not the smallest stellar cinders resulting from a supernova. Neutron stars are produced by the deaths of stars that are between about five and perhaps 30 times the mass of the Sun. If the star is more massive than this, the supernova produces a black hole.

Tiny neutron stars and invisible black holes are difficult to find in the vastness of space. One of the best ways is through X-ray astronomy. If a dwarf star is in orbit around a neutron star or black hole, the star may, as it ages, swell up to a giant and leak. Some gas falls on to the neutron star or black hole and is compressed by a strong force of gravity. Just as air compressed in a bicycle pump gets hot, so also this gas is heated – to temperatures in excess of a million degrees, causing it to emit X-rays. Of the first two interstellar X-ray sources discovered, Scorpius X-1 is thought to be a neutron star in such a double star system, while Cygnus X-1 is a black hole.

Very energetic versions of supernovae were discovered in a new manifestation in the 1960s by United States military satellites called Vela. They were monitoring whether the Soviet Union was carrying out nuclear tests, in the atmosphere or in space, which would violate the nuclear test ban treaty. Nuclear explosions produce gamma rays in short bursts, and to everybody's amazement the satellites immediately saw such bursts at a frequency of about once

Above *Nobel Laureate Antony Hewish and his student Jocelyn Bell inspect the curious tangle of wires on poles that they built, that constitutes the 4½-acre radio telescope with which pulsars were discovered.*

Below left *The Leviathan of Parsonstown in Ireland was a telescope made in 1845 with a six foot mirror. Pivoted at the base of the tube where the mirror was, the telescope barrel was slung between two masonry walls, hauled up and down in altitude and swung a little from side to side to track its targets for a period of time as they crossed the meridian. Access to the eyepiece at the upper end of the tube was by a precarious system of ladders and galleries.*

Below *The top trace of this strip chart was the first indication that the first pulsar discovered, CP1919, had pulses that were as regular as the timing marks below.*

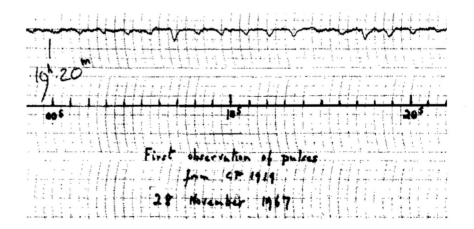

per day. This was far too many to be due to nuclear tests. Gamma Ray Bursts (GRBs) proved to come from random directions in space. Each burst lasted between a few thousandths of a second and several minutes.

Until 1997, astronomers did not know whether GRBs originated in some sort of explosions at the edge of our solar system, around our Galaxy, or far away. Two examples proved that the explosions occurred near the edge of the observable Universe. For their duration of a few seconds, the bursts had been over a million times brighter than their parent galaxy, the biggest bangs since the Big Bang.

The nature of some GRBs is still not clear, but astronomers think that the longer duration ones are especially energetic supernovae – sometimes called hypernovae. Taking place when the cores are "naked" – there is no material around them to cloak the gamma rays – hypernovae are the cores of stars collapsing to form black holes.

Above *GRB 080319B in 2008 was the most powerful gamma-ray burst ever detected. It produced a temporary star that was bright enough to be seen by the naked eye, a bright glow of X-ray emission (orange colour) and came from a galaxy at a distance of 8 billion light years.*

Insert above *Supernova 1994D, the star lower left, occurred in 1994 in the outskirts of the galaxy NGC 4526, a spiral galaxy tilted so that the dusty nearer side is silhouetted against the stars of its central regions.*

BLACK HOLES

Black holes were first postulated in 1783 by John Michell (1724–93) and in 1796 by Pierre-Simon Laplace (1749–1827). They are objects, perhaps stars, that are both small and massive, to such an extent that the gravity on their surface is very high, and nothing, not even light, can escape from them. Do black holes have real surfaces? Perhaps not, but surrounding a black hole is a surface called the event horizon from within which light cannot escape. If anything occurs within this surface, we can never know about it, which is the reason for the name "event horizon". Michell and Laplace put forward the concept of black holes, but it was the German mathematician Karl Schwarzschild (1873–1916) who, more than a century later in 1915, gave the modern formulation of black holes in terms of Albert Einstein's theory of General Relativity.

Left *The picture that gave the Crab Nebula its name, as drawn using the giant telescope in Ireland, at Birr Castle, and looking not much like the Hubble Telescope view right.*

Right *The Crab Nebula. A web of yellow and red filaments, pieces of the body of the star that exploded, enclose a hollow within which shines a pale blue light, generated by energetic electrons made by the pulsar at its centre.*

GRB970228 AND GRB 970508

Gamma-ray satellites like NASA's Gamma Ray Observatory (GRO) could not pinpoint the position of the bursters accurately enough for other telescopes to determine where exactly they were. Moreover, GRBs were over so quickly that, by the time a satellite had detected one and told ground controllers about it, it was over before other telescopes could train on it. A three-way coordination between a gamma-ray telescope, an X-ray telescope and optical telescopes solved this problem. GRO detected a burst called GRB970228 (the number is the date of the burst). Within an hour, an Italian X-ray satellite called BeppoSAX had interrupted its observing programme and slewed to the right direction, finding a new source of X-rays in the right place and measuring its position more accurately. On the phone immediately to Dutch colleagues at the observatory on La Palma in the Canary Islands, the X-ray astronomers asked them to search for something in this position. They found a new, faint source of light that was fading rapidly. The Hubble Space Telescope was reprogrammed to join in the examination, and, with its greater clarity, could see that the source of light was in a distant galaxy, about 8 billion light years away. A few months later another similar case occurred, GRB970508. The Keck Telescope on Hawaii measured the distance of this galaxy – it proved to be more than 6 billion light years from Earth.

The origin of the chemical elements

The nuclear reactions that generate the energy of stars change one chemical element into another. When stars are born, they consist primarily of hydrogen. In dwarfs, hydrogen changes into helium. In giants, the helium changes into carbon. In very massive stars that make supergiants, the carbon changes into oxygen, magnesium and neon, and then into elements like silicon and aluminium, perhaps transforming as far as nickel and iron. All the hydrogen in the core of the star gets used up, but the star has partly-processed material in layers around the core, structured like an onion, with unprocessed hydrogen in the outermost layer. This is left over because it has never been in a zone in the star hot and dense enough to change it.

Not all chemical elements have their origin in this way. There are other sites in astronomy where nuclear reactions take place that produce some of the less common elements. Some of these lie on the violent surfaces of certain energetic, stormy stars, but the most extreme conditions are found in supernova explosions – creating the new elements requires the sucking up of some of the energy of the supernova, and there is plenty of it. During these explosions, some of the elements of the material of the exploding star – helium, carbon, oxygen, iron and all the rest – is transformed further. One of the elements that is made this way is gold – when you look at that nice gold jewellery you can continue to remember the pleasant moment when you were given it, but you could also imagine its origin in a supernova explosion of the distant past, billions of years before that.

Astronomers actually witnessed the creation of elements by a supernova in 1987 in the galaxy that neighbours ours, the Large Magellanic Cloud. Some of the nickel in the star was transformed into cobalt, in a form that is radioactive and decays over about a year to iron. This form of cobalt creates gamma rays, which were picked up by space satellites, including one called Solar Max. As its name implies, the satellite was studying the Sun, pointing its telescopes away from the Large Magellanic Cloud, but the gamma rays from the supernova were so energetic that they penetrated the body of the satellite and activated its detectors anyway.

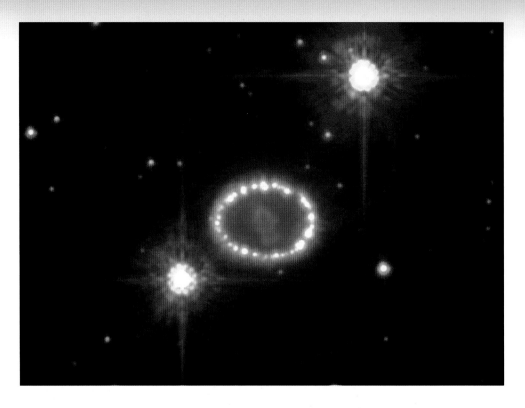

Above *The bead-like points of light forming a halo are clouds of gas lit up now that fragments of the explosion of a supernova of 1987 are colliding with them. This is the way that elements made in the bodies of stars are ejected into and mix with the gas surrounding the star. The mixture is destined to become future planetary systems, whose composition will have been enriched by elements formed and distributed in this way.*

Left *William Fowler, physicist and Nobel Laureate.*

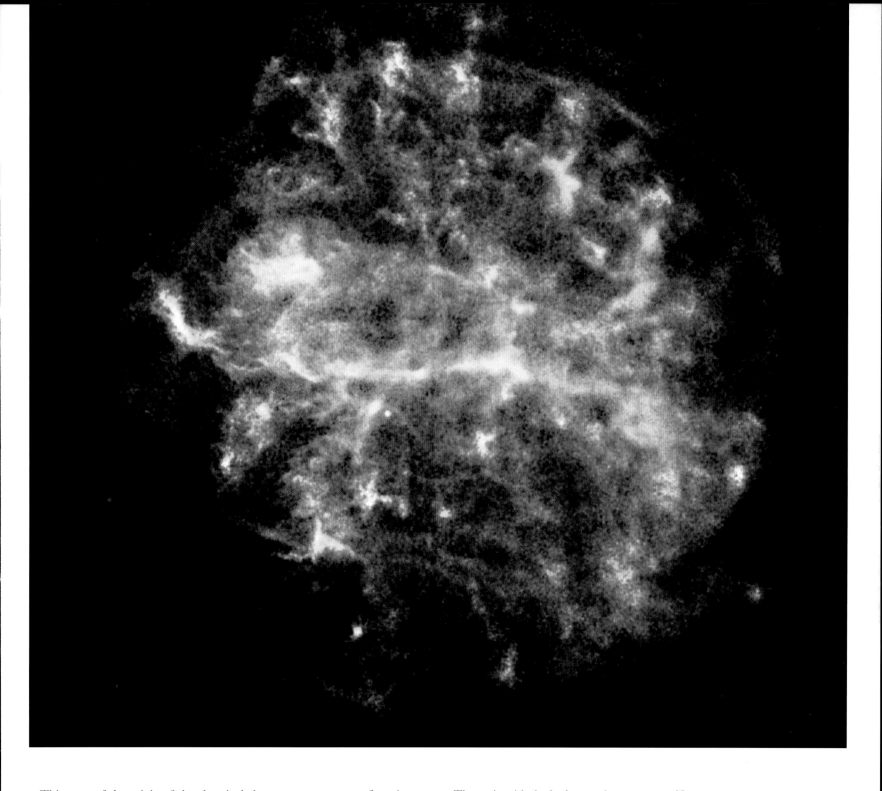

This story of the origin of the chemical elements was put together first in the 1950s by the Anglo-American team of Geoffrey Burbidge (1925–2010), Margaret Burbidge (1919–), William Fowler (1911–95) and Fred Hoyle (1915–2001) in a detailed, epoch-making scientific paper known by the initials of its authors as B^{2}FH ("B-squared, F, H"); some of the work was independently done by Harvard astrophysicist Alistair Cameron (1925–2005). This cross-disciplinary work attracted an array of prizes, including a Nobel Prize for Fowler in 1983.

When a massive star loses its material into space as it dies, whether by generating a planetary nebula, or by exploding as a supernova, the elements in the outer layers are flung into space. They mix with the hydrogen there, enriching it with the newly made elements. Any new stars that form out of this hydrogen in the future will have these new elements mixed in. And if the star spins off a cloud of material into a disc at the start of its life this material can condense into planets. The fundamental reason why the Earth has a rocky crust, made of minerals with, predominantly, carbon, oxygen, aluminium and silicon, and a core of iron and nickel, is that these elements were made in abundance in stars that exploded before the Sun was born. Our planetary environment – including the material of which we ourselves are made – has been created by stars from generations that pre-dated the birth of the Sun.

Above *A supernova remnant known only by its catalogue number G292.0+1.8, imaged in X-rays by the telescope on the satellite Chandra. The star-like image surrounded by a pink-blue light is a neutron star, the stellar cinder made by the explosion of the progenitor star about 1,600 years ago. The large spherical cloud of gas that surrounds the explosion point is the scattered body of the progenitor and is rich in oxygen, the element having been made by nuclear fusion of carbon in the progenitor.*

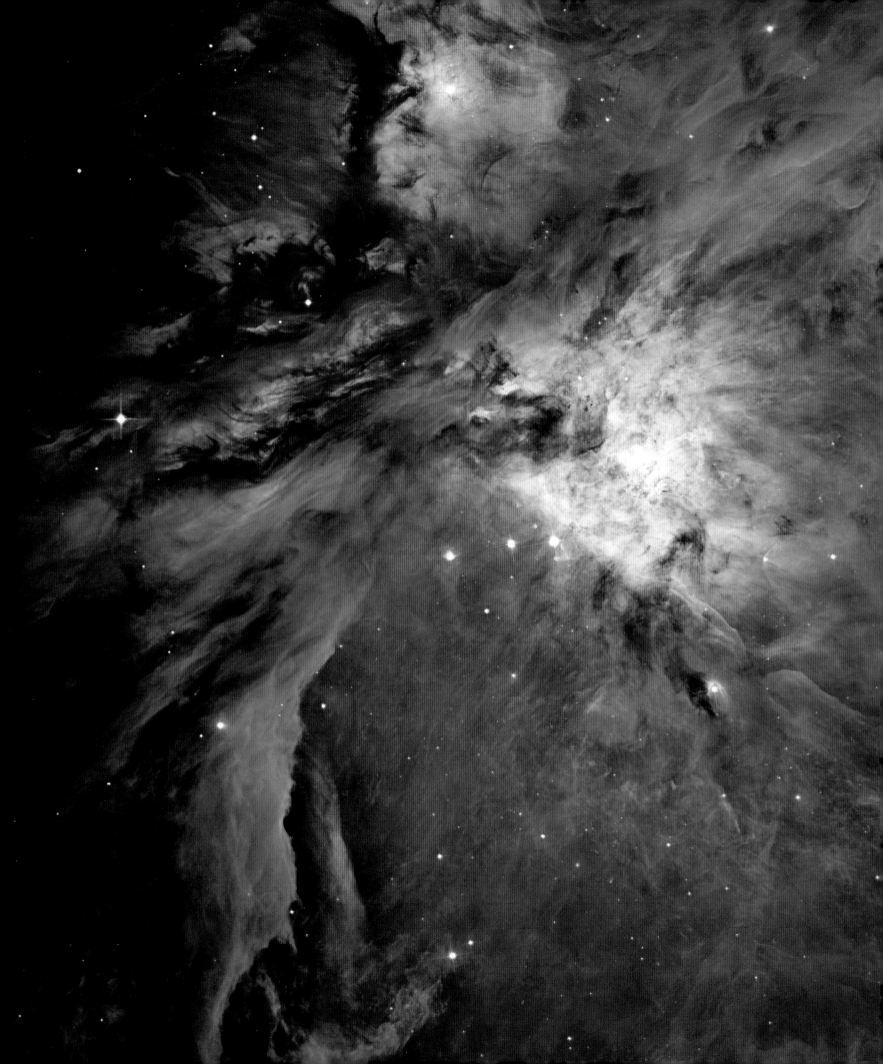

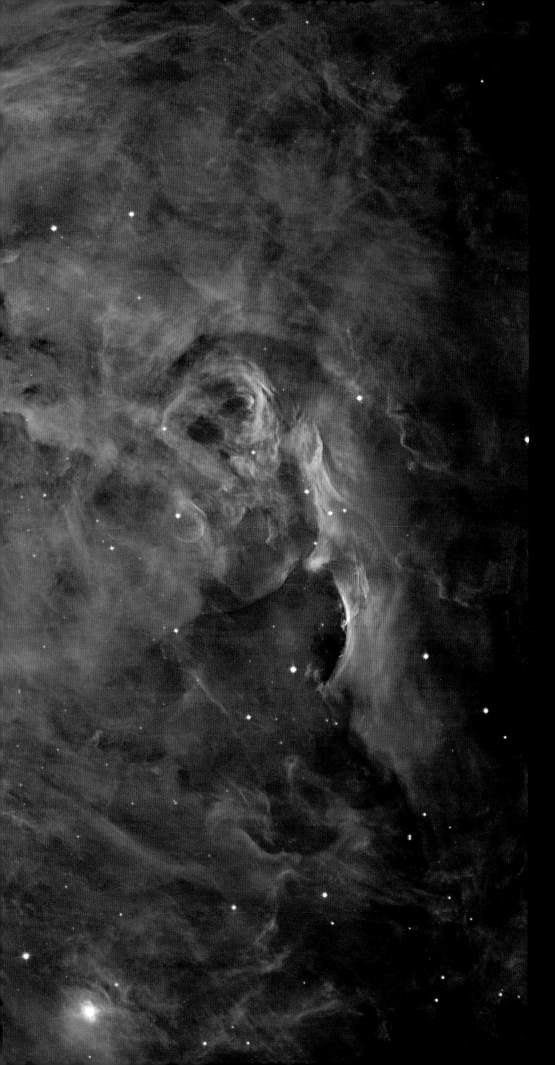

NEBULAE

Stars form in clusters out of clouds in interstellar space. A cluster of stars contains stars of different masses, including some that are massive, bright and hot. Such stars radiate ultraviolet light that heats the surrounding nebula and causes it to glow. The bright stars of Orion are of this sort and one of them, the central star in the Sword, is surrounded by such a nebula. It was the first nebula discovered, by the French lawyer and amateur astronomer, Nicholas-Claude Fabri de Peiresc (1580–1637) in 1610, using a new telescope bought for him by his patron Guillaume du Vair. The Orion Nebula was independently discovered the next year by Jesuit astronomer Johann-Baptist Cysat (1588–1657).

At only 1,500 light years away, the Orion Nebula is the closest and thus the most studied nebula in the sky. Viewed in optical telescopes, like the Hubble Space Telescope, the Orion Nebula is the illuminated inner surface of a dish-like hollow in the surface of a much larger, dark interstellar cloud, like a bite in an apple. Poised in the dish are four stars known as the Trapezium; indeed, winds from these four stars have probably carved out the dish, returning back to the nebula some of the material that created them. The Trapezium stars are just the nearest of a cluster of hundreds of new-born stars that lie inside the dark cloud. These stars reveal themselves as sources of infrared radiation, mapped in detail by space satellites with infrared telescopes on board, most recently by NASA's Spitzer satellite, launched in 2003.

Left *The Orion Nebula in a picture mosaicked together from 520 individual Hubble Telescope pictures. The Trapezium of four stars lies in a bowl centred on the brightest part of the nebula. A sheet of obscuring dust curls over the nebula from the upper left corner like the fingers of a dead hand.*

The birth of stars and planets

Left *Pierre-Simon Laplace, philosopher and mathematician.*

In 1796, the French astronomer Pierre-Simon Laplace (1749–1827), proved mathematically that the shape of the Solar System, in which the planets all orbit the same way round the Sun in a flat disc, is the way that the Solar System was formed – it has stayed that way ever since birth. His proof supported an idea put forward by the Swedish scientist Emanuel Swedenborg (1688–1772) in 1734 and the Prussian philosopher Immanuel Kant (1724–1804) in 1755, that the planets condensed out of a flat nebula whirling around the Sun. This idea became known as the Nebular Hypothesis

Laplace thought that the so-called "planetary nebulae" found by William Herschel were examples of proto-planetary systems. In fact, as we saw earlier, they were not new stars at all – it was two centuries later, only one generation ago, that the first proto-planetary systems were actually discovered.

The first was found in 1966 in the Orion Nebula by Cal Tech astronomers Eric Becklin (1940–) and Gerry Neugebauer (1932–). Invisible to optical astronomers, the "BN Object" (from the discoverers' initials) is a strong source of infrared radiation. A dust cloud shrouds a newly formed star and is heated up, giving off the infrared. The InfraRed Astronomy Satellite (IRAS) discovered further proto-planetary systems in 1983, including dust discs orbiting the stars Zeta Leporis, Vega and Beta Pictoris. In the last star, a small planet orbits inside the disc.

The first direct images of proto-planetary systems were made in 1992 by Robert O'Dell of Rice University with the Hubble Space Telescope. They were dust discs silhouetted against the luminous background of the Orion Nebula, each with a central star, just as the Nebular Hypothesis imagined. The objects were termed "proplyds" – a contraction of "proto-planetary discs".

Let us follow the way that the Sun and Solar System started, from their early moments in a contracting gas cloud, as first calculated by the Japanese astrophysicist Chushiro Hayashi (1920–2010) in 1960. The proto-star that became the Sun started as a collapsing sphere, but went through spasms in which it ejected a stellar wind in every direction and squirted jets of material from its poles. The sphere became flattened into a disc.

The proto-star took up a rapid spin as it contracted in the interstellar cloud – any slow rotation by the cloud was sped up as the cloud contracted, just as, in a final flourish at the end of her dance routine, an ice skater will spin faster as she brings her arms closer to her body. The star slowed by throwing off material into the disc of a planetary system.

The proto-planetary system was mostly composed of hydrogen and helium, but it also contained elements that had been made in old stars and exploded into the interstellar cloud. Some of these elements made dust grains – graphite grains and little diamonds from carbon, silicate grains (sand) from silicon and oxygen. Dust grains made in this way were the ones discovered by astronomers in the BN Object, and imaged by the Hubble Space Telescope. Some of the elements in the gas cloud connected into molecules which condensed as ice on the dust grains. When the proto-star switched on its nuclear reactions, it began to radiate energy,

Below *The star Beta Pictoris has been obscured by an obstruction in a camera at the centre of this picture so that its light does not swamp the fainter structure that surrounds the star. The faint structure is a disc of dust seen edge-on, in which there is a small planet. Our Solar System looked like this, once.*

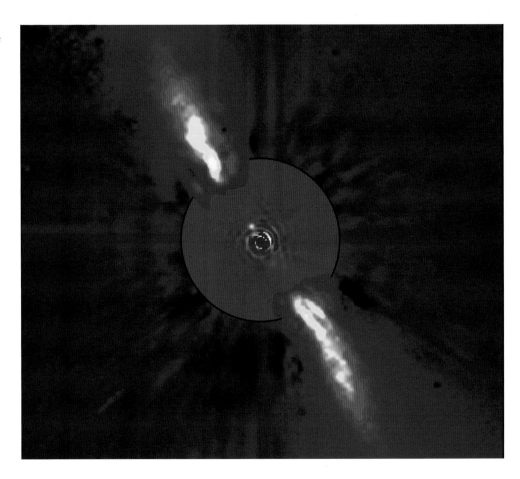

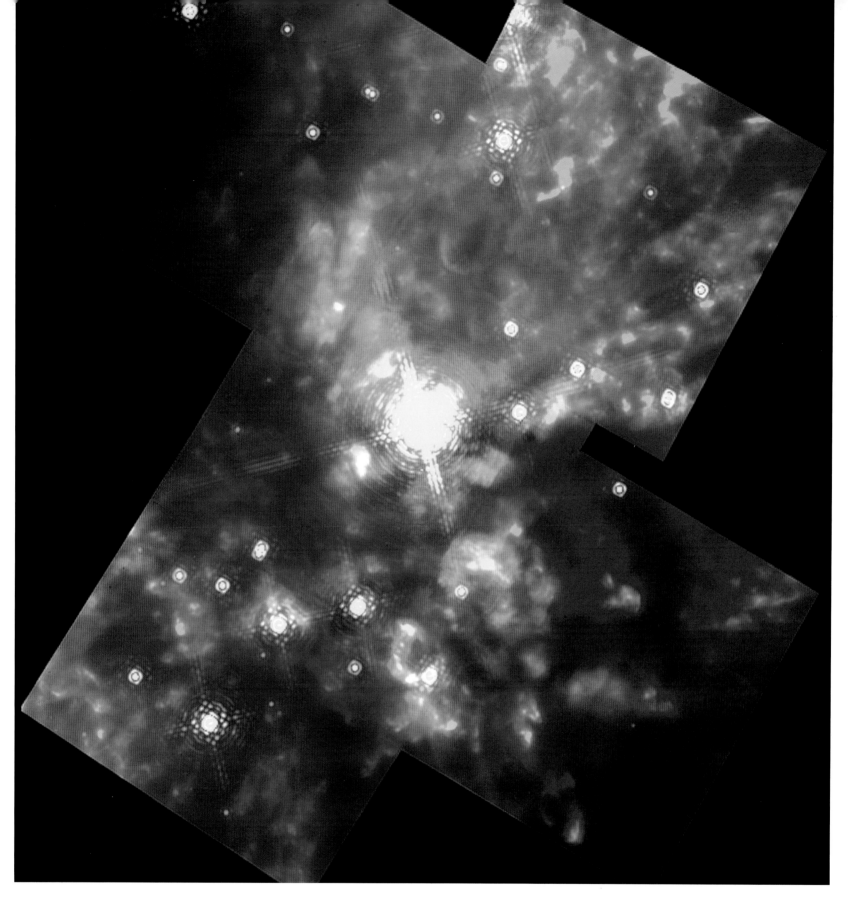

Above *The Hubble Space Telescope has infrared vision that provided a look through the dust that obscures the interior of the Orion Nebula, a region known as OMC-1. It is a chaotic, active region where stars are being born, the brightest being the BN Object, a very young, massive star. Blue "fingers" of hydrogen emission radiate across the image, outflowing material from the newly formed stars.*

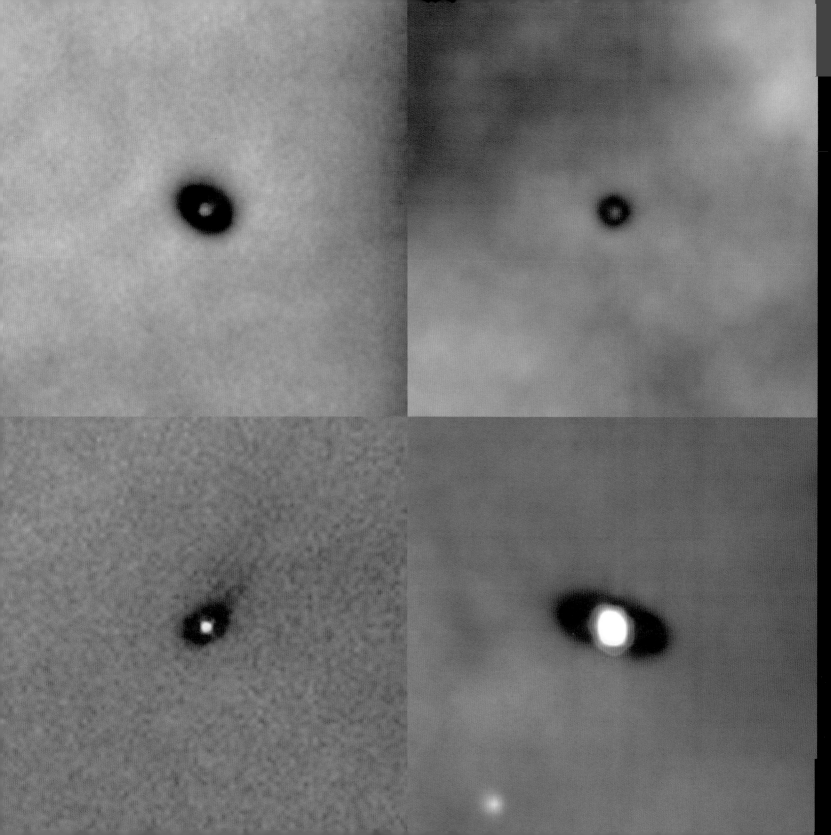

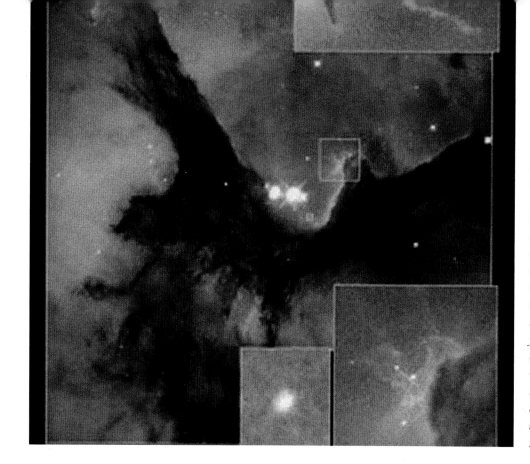

Left Three huge intersecting dark lanes of interstellar dust trisect the gas of the Trifid Nebula. Near their intersection is a group of recently-formed, massive, bright star, which have reduced the density of the gas and they produce a bright rim on the side of a lane facing the stars (inset bottom right). Not far away there are still signs that stars are still forming in the nebula. One very young star is still surrounded by a ring of gas and dust left over from its birth – a proplyd (inset at bottom). At the top there is a jet of material three-quarters of a light-year long being ejected from a very young, low-mass star.

Opposite The red glow in the centre of each disc is a young, newly formed star, roughly one million years old. The dark, dusty discs are silhouetted against the bright backdrop of the Orion nebula, and these discs, apparently protected against the scouring effects of the flow of gas in the nebula are likely to become planetary systems.

Right A chondrite meteorite, its outer skin blackened and bubbled by the heat of its fall to Earth through the atmosphere. Its interior, exposed when the meteorite was cut in half, is a conglomeration of "chondrules", amalgamated dust grains that were part of the solar nebula. This meteorite failed to become part of the planets, like Earth, which built up from chondrules like these.

melting the ices. It blew away gas and vapour in the inner, warmer parts of the solar nebula.

In the proto-planetary nebula, grains of dust then stuck together in solid lumps, which were massive and not easily destroyed or expelled. The lumps grew to kilometres in size. At this stage they are called "planetesimals".

The gravity of the planetesimals was high enough to attract one another – a large planetesimal attracted smaller ones, and grew bigger, so its gravity increased, so it grew bigger still. This process is called "accretion". It resulted in about 100 proto-planets in orbit in the inner Solar System. Accretion stopped when the proto-planets had fed on everything around them, but some of the small fragments that were left over fall to Earth from time to time. They are a particular kind of meteorite called a chondrite.

Material in the more distant part of the gas cloud was less affected by the heat of the star, so it is gas-rich and built up the gas-giant planets. The planet Jupiter grew the fastest and the largest amount.

What followed in the next billion years was a gigantic game of interplanetary billiards, according to calculations made in 2005 by Rodney Gomes, Hal Levison, Alessandro Morbidelli and Kleomenis Tsiganis, an international group of mathematicians centred on the Côte d'Azur Observatory in Nice, France. Their calculations are therefore known as the Nice Model of the history of the Solar System.

Jupiter's gravity stirred up the orbits of the inner proto-planets. It stopped the formation of a single planet in the region inside its own orbit, towards Mars – some planetesimals have been left there as asteroids. Some planetesimals jay-walked across the main flow of revolution of the others, and collisions were inevitable. Various impacts aggregated some of these planetesimals into the terrestrial planets – Mercury, Venus, Earth/Moon, Mars. Some impacts shattered other planetesimals – the pieces became the most common of the asteroids. One impact produced only two pieces: it produced a "twin planet", the Earth and the Moon. The fragments of shattered planetesimals rained down to make the large craters on the Moon and Mercury – and the Earth, although the weather has eroded these.

In the chaos of this time, many smaller planetesimals were ejected into space and lost to the Solar System. Those that were not quite lost took up orbits in distant parts of the Solar System, called the Kuiper Belt *(see page 48)* and the Oort Cloud *(see page 49)*. We see these planetesimals today as comets and Kuiper Belt asteroids like the dwarf planet Pluto.

Exploration of the planets

The space age began in 1957 with the launch of the Russian Sputnik satellite. The next year, the first successfully launched American satellites, Explorer I and III carried relatively simple scientific equipment made by James Van Allen (1914–2006) to explore the near-space environment of the Earth. These satellites discovered that energetic, electrically charged solar particles are trapped in the magnetic field of Earth, the so-called Van Allen Belts. This was the first major scientific discovery of the space age.

The Moon is the nearest celestial body to the Earth and its exploration is furthest advanced, from the first approaches in 1959 by the Soviet Luna spacecraft, through to the US Apollo programme, with the first manned lunar landing by Apollo 11 in 1969.

Below *In July 1969 Apollo 11 astronaut Buzz Aldrin, one of the first two men on the Moon, stands next to a seismometer, which measured "moonquakes". In the background, beyond moon-boot footprints in the dust, is Eagle, the lunar lander. The seismometer and the lower part of the lunar lander are still there on the lunar surface.*

Opposite *The Galileo spacecraft imaged our own Moon in 1992, on its way to Jupiter, and showed well the variety of its craters, from large, eroded, old craters flooded with uniform grey lava, crossed by rays from some of the smaller, more recent craters with craggy walls.*

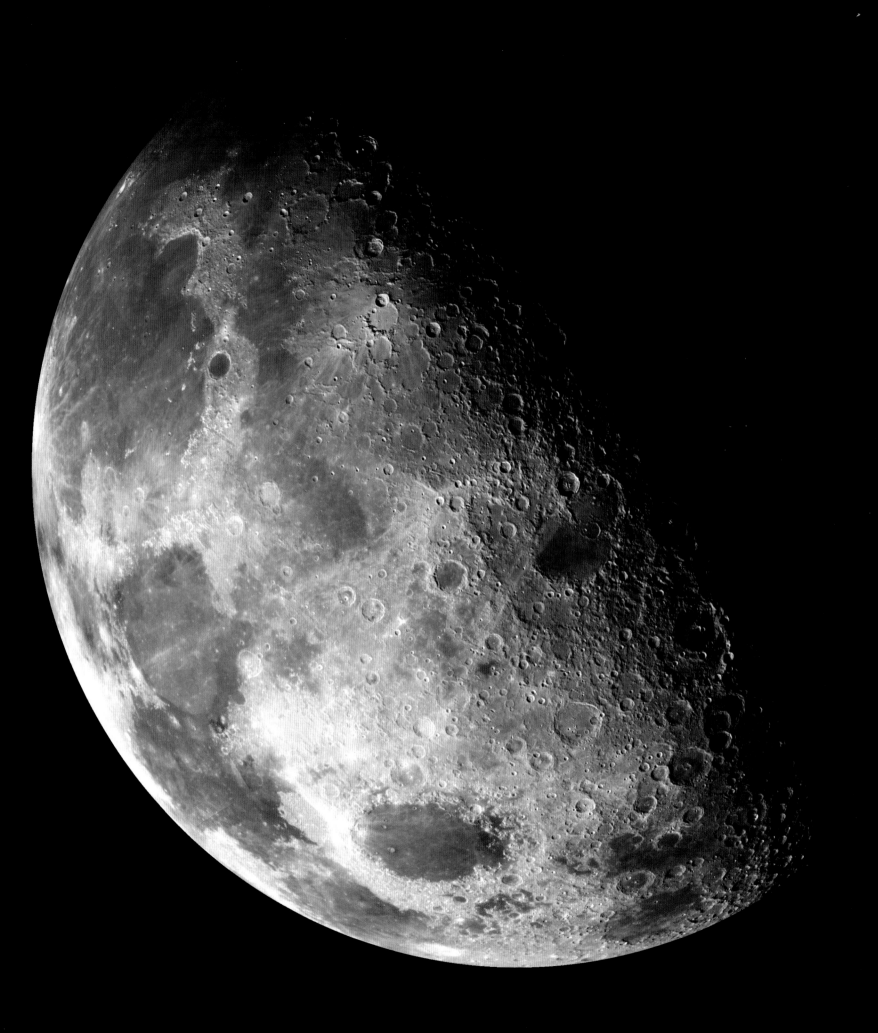

ROBOTIC EXPLORATION

Scientists across the space agencies coordinate their planetary exploration programmes in bodies like the International Mars Exploration Working Group. A programme typically starts with a flyby – a quick look as a spacecraft zooms past – and then makes sustained investigations remotely from orbit. Probes are landed through the atmosphere and on to the planet's surface; parachutes slow their free fall (if the planet has an atmosphere, like Venus, Mars, Jupiter and Titan) or up-thrusters (as for landing on the Moon). Landers are stationary on the surface but rovers are free to move until they get stuck, or run out of motive power when the solar panels get dusty. The exploration programme may return samples back to Earth for close study (so far achieved only for the Moon, a comet and an asteroid), with manned exploration and, in future, colonization the climax of the programme.

Above *The Sojourner rover on the Mars Pathfinder lander spacecraft in July 1997 just before it descended to the surface of Mars over the deflated airbags that protected it during the impact. On the horizon the "Twin Peaks" are about 2 km (1.25m) away.*

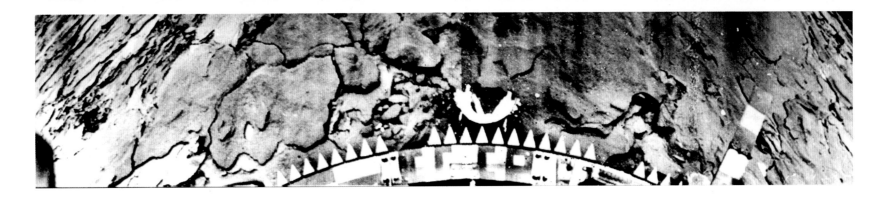

Opposite *The surface of Venus as viewed by the camera of the Soviet Venera 14 lander in March 1982 is weathered, layers of rocks. The lander carried a spring-loaded arm to test how much the surface would compress but unfortunately the ejected lens-cap of the camera (centre of the top picture) fell exactly where the probe tried to press the ground. Nevertheless the lander established that the soil was tholeiitic basalt (the most common type of volcanic rock on Earth, found in abundance on Earth's mid-ocean ridges).*

Venus is the next closest celestial body. In 1962, Mariner 2 executed a flyby of the planet Venus, with the first landing by Venera spacecraft in the 1970s revealing its rocky, lava-strewn surface. The first flyby of Mars was by Mariner 4 in 1965. Ten years later two Viking Landers revealed close-up the arid, rocky surface of Mars. Their orbiters found evidence that regions of the surface had at one time been awash with water. Mars Pathfinder was the first successful Mars rover vehicle in 1997.

The more remote the planet the more difficult it is to get there and operate. Sunlight is less intense in the distant Solar System, but to counteract this spacecraft can use radioactive thermoelectric generators (RTGs) instead of solar panels to store energy and make electricity. Voyager 1 and Voyager 2 were launched on so-called "Grand Tours" in 1977 to travel past Jupiter, Saturn, Uranus and Neptune, and their satellites. They used the technique of gravity assists – approaching a planet in such a way that the spacecraft picks up speed from its gravitational pull in a slingshot effect.

Pluto remains unexplored by any spacecraft, but the New Horizons mission was launched in 2006 and will reach the planet in 2015.

Above *There is a vast region of sand dunes around the north polar cap of Mars, which in the winter is covered and solidified by a layer of carbon dioxide ice. The ice is basically white, but is tinted orange by the red dust of Martian soil blown over its surface by planet-wide dust-storms originating near the equator during the winter. In the Martian spring the Sun warms and evaporates the ice. Sand particles, now unbound and freed on the slopes of the dunes, cascade down their sides. The landslips show as dark streaks that have ploughed up the surface frost layers. Left of centre, a surge of material falling just before the image was taken by NASA's Mars Reconnaissance Orbiter has kicked up a small cloud of billowing orange dust. The red-orange colour of diffuse areas of ice surrounding the ends of streaks of material elsewhere in the picture suggests that dust has settled on the ice after similar events.*

The inner planets

Two planets revolve around the Sun inside the Earth's orbit, Venus and Mercury. Because they orbit close to the Sun, they can be seen only in the twilight, and in classical times (before the fourth century BC) each planet was thought to be two different planets, depending on whether it was seen in the morning or the evening. Mercury was known as Apollo and Hermes, Venus as Eosphoros and Hesperos (the morning and evening stars respectively).

Mercury is somewhat larger than the Moon, and has a similar, heavily cratered surface – the result of intense bombardment by asteroids early in the history of the Solar System – and no atmosphere. It is so close to the Sun that its surface temperature is 430 °C (806 °F) at the equator. However, the lack of atmosphere means that at the poles it is extremely cold (-183 °C; -297 °F) inside high-walled craters, where sunlight never reaches.

Mercury is small and difficult to observe from Earth, because it is so near to the Sun. It is not easy to study with spacecraft, either, since the Sun causes approaching spacecraft to accelerate and they tend to overshoot. Moreover, the Sun's heat interferes with the operation of the spacecraft's equipment. The first problem was solved by the Italian scientist Giuseppe ("Bepi") Colombo

Right *Venus is an evening or morning star. Mercury is closer to the Sun and generally lies even closer to the rising or setting Sun. Here the two planets struggle to display their natural beauty above the man-made glory of the brightly-lit church of Notre Dame in Paris*

Below *The Venera 13 spacecraft had landed a few days before Venera 14, 1000 km from it. It took colour pictures of the surface (the left and right halves of a picture are in the two frames reproduced here). The rock here was leucite, a volcanic rock that breaks down quickly and is therefore of relatively recent origin.*

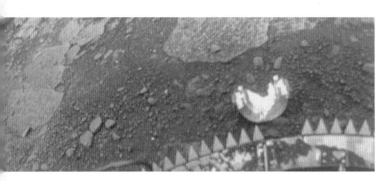

(1920–84), who calculated an orbit for the NASA space-probe Mariner 10 that used Venus to pull the spacecraft into an orbit that made it execute repeated flybys. A future ESA spacecraft mission to Mercury is being called BepiColombo in his honour. A similar idea has been used to insert a NASA space probe called MErcury Surface, Space ENvironment, GEochemistry and Ranging (MESSENGER) towards an orbit around Mercury for a year in 2011–12.

Mariner 10 and MESSENGER together have mapped the whole of Mercury's surface. Its largest meteor crater is Caloris Basin, 1,550 kilometres (965 miles) in diameter, one of the largest in the Solar System. At the opposite end of the diameter through Caloris Basin is a large region of unusual, hilly, so-called "Weird Terrain", caused by shock waves from the impact that travelled around the planet, and converged on the other side.

Venus is very similar in size to our own planet and is sometimes called Earth's twin, but it is an evil twin, a planet with a fiery, volcanic surface, a heavy, dense atmosphere of carbon dioxide and a foul climate of hot, sulphuric acid rain and 320 km/h (200 mph) high-altitude winds. The sulphuric acid droplets produce

Right *Maat Mons is an 8 km (5 miles) high volcano on Venus, reconstructed by computer in this perspective view from topographic maps made by the Magellan spacecraft. Its lava flows extend several hundred kilometres from its summit.*

Below *The Messenger spacecraft took this colour picture of Mercury in 2008. The surface is peppered with craters, some of the more recent, bright, white craters with rays of ejected material, like craters on the Moon. The large circular light-coloured area in the upper right of the image is the Caloris basin.*

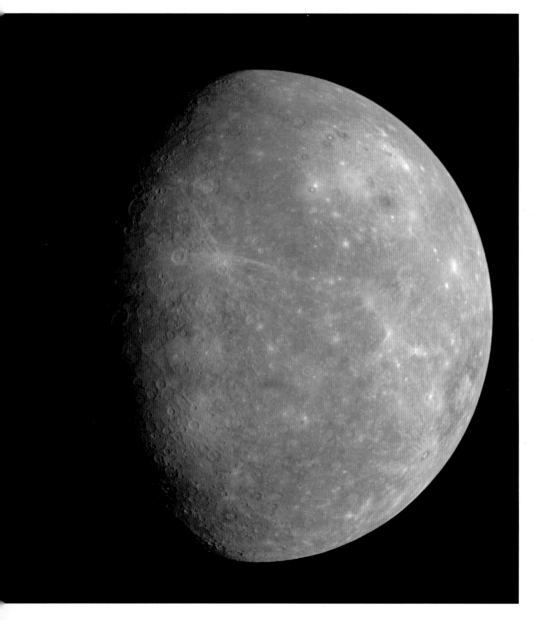

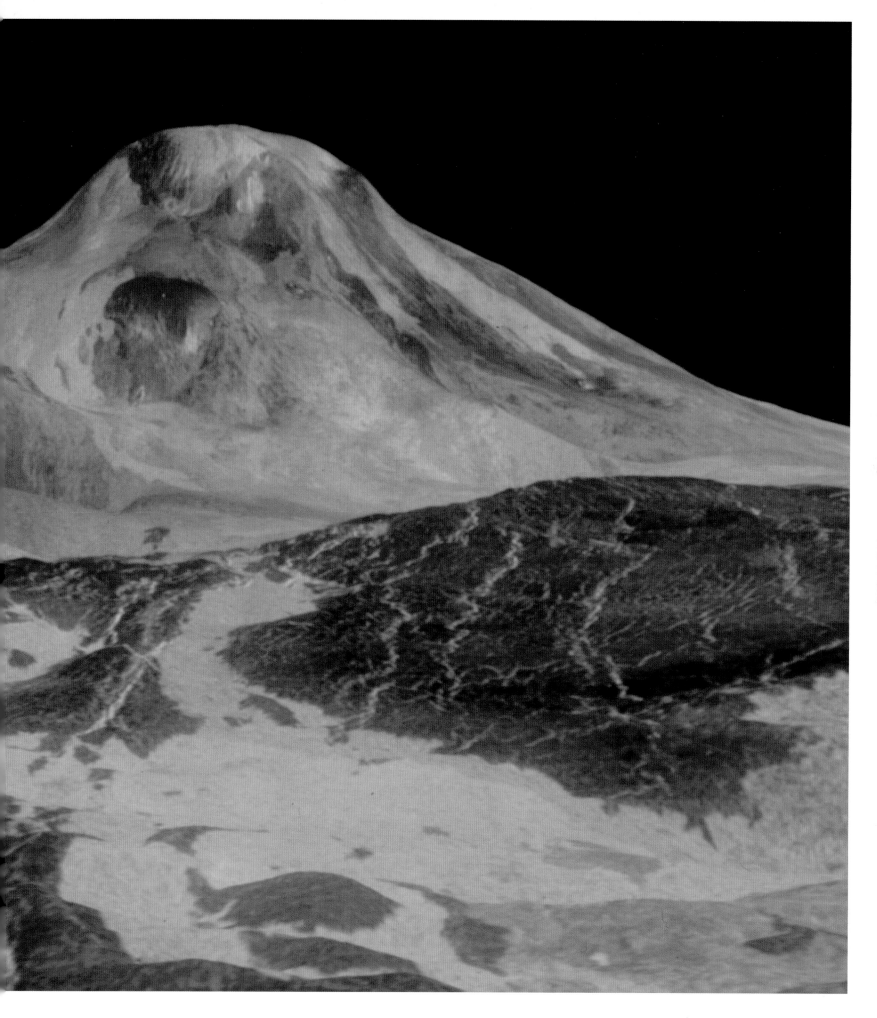

opaque clouds, and so in a telescope the planet is almost featureless – little or nothing can be seen of the surface.

Most of what we know about the planet is the result of the space exploration programme, which began in 1962 with the United States Mariner 2 flyby mission, the first visit by a spacecraft to another planet. Landings by the Russian Venera spacecraft followed; the conditions were so hostile they operated for minutes rather than days.

The United States Magellan spacecraft remained in orbit around Venus for over four years. Its radar instrument penetrated the clouds and mapped almost the entire surface, showing its volcanic features – lava plains, lava channels and volcanoes, of which there are over 100 times as many large ones as on Earth. There are relatively few meteor craters on Venus, because its surface is young – perhaps half a billion years ago there was considerable volcanic activity which resurfaced the planet.

ESA's Venus Express mission arrived to orbit Venus in 2006 and has been concentrating on studying Venus's atmosphere. Its atmosphere was discovered in 1761 by the Russian astronomer Mikhail Lomonosov, backlit by the Sun during a transit of the Sun. Venus's atmosphere is composed of dense carbon dioxide with a very strong greenhouse effect which retains that part of the Sun's heat that reaches the surface: the temperature there is over 460 °C (860 °F). Venus once had surface oceans, which evaporated due to the greenhouse gas effect of carbon dioxide. The water vapour this released augmented the greenhouse effect, increased the temperature and caused further greenhouse gases to be released. The resultant runaway greenhouse effect is the reason for the nature of Venus's climate today.

Opposite *Caloris Basin on Mercury in a mosaic of images from the spacecraft Mariner 10. There are concentric arcs surrounding this large meteor crater, which is centred in the dark side of the planet to the left of the picture.*

Above right *As Venus drifted on to the disc of the Sun during the transit of 1874 as viewed from Sydney, Australia, its atmosphere refracted sunlight to complete the circle of its silhouette.*

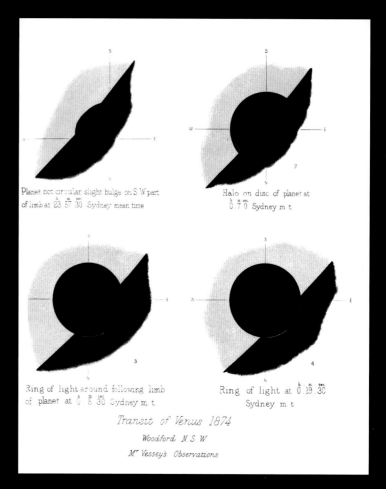

INSPIRING ORBITS

From time to time (occurring on two occasions every century, separated by approximately eight years), Venus passes across the face of the Sun (a "transit"). The event can be used to determine the distance to the Sun, an essential piece of data for the calculation of ephemerides (tables for the prediction of the positions of the planets for navigational purposes). International expeditions were mustered to measure the transits of 1761 and 1769 from multiple locations, including Siberia, Norway, Newfoundland, Madagascar, the Cape of Good Hope – where observations were made by geodesists Jeremiah Dixon (1733–79) and Charles Mason (1728–86) – the surveyors of the Mason-Dixon Line in pre-Independence America – Hudson Bay, Baja California (then under Spanish control) and Tahiti. The expedition to Tahiti was led by Captain Cook, on his first circumnavigation of the globe during which he discovered Australia.

Mercury's orbit played a crucial part in the development of Albert Einstein's (1879–1955) General Theory of Relativity. Einstein's theory was more accurate than that of Newton, especially when describing the behaviour of objects in strong gravitational fields, such as near the Sun. This strong gravitational pull makes Mercury's eccentric orbit precess, i.e. the long axis of the orbit is not static in space but drifts in the direction of the orbit. This drift had already been observed in the late nineteenth century and was interpreted by Urbain Leverrier (1811–77) to mean there is a planet even closer to the Sun than Mercury – provisionally named Vulcan – which was perturbing Mercury's orbit. No such planet was ever found, and for another 100 years the precession remained a great puzzle. When Einstein discovered in 1915 that his theory neatly and unexpectedly solved the problem, he was overjoyed and encouraged to reveal his theory to the world.

The Earth, the Moon and Mars

Earth

The Earth's history as a planet began with Copernicus in 1543, who recognized that it orbited the Sun, and was confirmed in the late 1960s when the Apollo astronauts first saw the complete Earth from outside and confirmed its nature as a world like other planets. The Earth was seen first in its entirety as a planet in 1968. After the *Apollo 8* astronauts, Frank Borman, Jim Lovell and Bill Anders, had passed behind the far side of the Moon, they photographed Earth rising above the Moon's surface. The blue, white and brown sphere of the Earth, with its oceans, clouds and continents, contrasted vividly with the monotonous, grey, dusty landscape of the Moon. The photographs became iconic images of the limitations of our planet.

The Earth is unique among the planets due to having extensive tectonic activity, including volcanoes and earthquakes. In 1912 the German geophysicist Arnold Wegener (1880–1930) noted that the continents fit together like the pieces of a jigsaw. To general ridicule, he hypothesized that the continents had once been part of a single landmass and are drifting apart. In the 1950s the Mid-Atlantic Ridge of sub-oceanic mountains and volcanoes was discovered, marking the line at which material was upwelling into the ocean, driving the continents of North and South America and of Europe and Africa apart. They "floated" on the Earth's mantle (the pliant layer between the solid crust and the liquid core of our planet). This happens because, proportionately, the Earth has the largest liquid core of any of the planets: its tectonic activity is related to the unique origin of this large core.

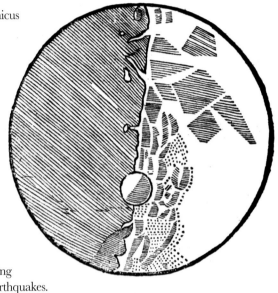
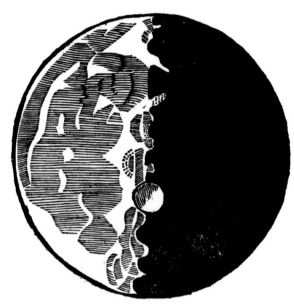

The Moon and its origin

With his telescope in 1610, Galileo saw the grey lunar plains and the lighter areas of the lunar highlands, with mountain ranges. Across all was a peppering of lunar craters. For 300 years they were thought to be volcanoes. In 1903 US mining engineer Daniel Barringer (1860–1929) suggested that the Arizona Crater was a meteoric impact crater, a hypothesis confirmed in 1960 by geologist Eugene Shoemaker (1928–97) with the key discovery there of rare forms of silica created from severely impacted quartz. In six missions to the Moon (1969–72) the Apollo astronauts brought 300 kg (661 lb) of lunar rock back to Earth for analysis. They have been pounded by meteor impacts. The most startling thing is that their composition is identical in key ways to that of rocks found on Earth, which suggests that the Moon and the Earth were formed in the same process.

In 1946, the US geologist Reginald Daly (1871–1957) suggested that the Moon was formed by something impacting on the Earth. This hypothesis was reinvigorated in modern form during discussion at a conference in Hawaii in 1974. The suggestion is that the Moon was formed by the tangential collision of a body, now called Theia, the size of Mars, with the proto-Earth. This gave the Earth an extra spin at the same time that a cloud of mantle material was splashed into orbit round the Earth, material that coalesced to build up the Moon. The liquid cores of Theia and the proto-Earth merged and stayed behind as a double sized core in what became the Earth. Earthquakes have their ultimate origin in this chance process.

Above *In his book* Sidereus Nuncius *Galileo illustrated the First and Last Quarter Moons, showing the crater Albategnius unrealistically large and dramatic on the line between illuminated and dark side of the Moon. These are woodcuts from a pirated copy of Galileo's book (Frankfurt 1610).*

Opposite *The Moon's sterile, dry grey sphere contrasts with the vibrant blue, white and green sphere of the watery Earth, a picture symbolic of our planet's finite resources.*

Above *A historic globe of Mars (early twentieth century) shows the bright, white south polar cap, surrounded by a dark fringe then thought to be wave of vegetable growth stimulated by melting snow, and the network of* canali — *then thought to be rivers or canals that distributed water over the planet.*

Mars

Mars is a red colour and in astrology is associated with the god of war – the planet has given its name to the adjective "martial". It has always had a threatening image.

Galileo, the first person to view Mars through a telescope in 1610, could see that it had a disc, but he could discern no features. Using improved telescopes, the Dutch astronomer Christiaan Huygens (1629–1695) detected in 1659 a shadowy grey-green triangular feature known now as Syrtis Major. The Italian-born French astronomer Giovanni Domenico Cassini (1625–1712) saw in 1666 a white cap over the south pole of Mars. In 1704, the French-Italian astronomer Jacques Philippe Maraldi (1665–1729) showed that the size of the polar caps varied throughout the martian seasons, as if they were ice caps like Earth's. By 1781, William Herschel *(see page 44)* was articulating the growing realization that Mars had a "considerable but moderate atmosphere, so that its inhabitants probably enjoy a situation in many respects similar to ours".

Taking advantage of times when Mars was closer to Earth than usual the Vatican Observatory astronomer Angelo Secchi (1818–78) drew detailed representations of Mars in 1869 which included two dark linear features on the surface that he referred to as *canali*, which is Italian for "channels" but which, literally translated into in English, suggests artificial construction.

In 1877 the Italian astronomer Giovanni Schiaparelli (1835–1910) produced the first detailed map of Mars, which showed numerous *canali*, to which he gave names of famous rivers on Earth. French astronomer Camille Flammarion (1842–1925) wrote that these channels resembled man-made canals, which could be used to redistribute water across the planet. The US businessman Percival Lowell founded the Lowell observatory to investigate Mars in 1894 and published highly successful books about life on Mars, including maps with over-detailed networks of canals. All of this encouraged the idea that Mars was a dying planet, inhabited by advanced beings, and served as the foundation of H.G. Wells' novel *War of the Worlds* (1898), and the subsequent spin-off movies, plays and concept records.

Below *Phobos is locked in its orbit around Mars so that the same place is always in front. It has been scored with parallel chasms that run from there, some of them running through older, more eroded craters, some of them interrupted by sharper, newer ones. They are apparently caused by glancing blows by meteorites overtaken by the satellite.*

Above Astronomical Observations: Mars. *Donato Creti's paintings of 1711 show gallants and a sage inspecting the skies, and what they can see. This one shows Mars as a slightly gibbous, featureless sphere.*

THE MOONS OF MARS: PHOBOS AND DEIMOS

In 1877, using a telescope at the US Naval Observatory, the American astronomer Asaph Hall (1829–1907) discovered the two moons of Mars, named Phobos and Deimos (Fear and Dread, as summoned by Ares, the God of War, in *The Iliad*). As seen in close up pictures by spacecraft sent to explore the Martian system, the most recent and most detailed by the ESA Mars Express mission in 2008, both moons look like asteroids, irregularly shaped like potatoes and pitted with craters caused by meteor impacts. One guess is that they are indeed asteroids that have been captured as they orbited too near to Mars; however there is no satisfactory calculation for how exactly this might have happened.

Space exploration of Mars

The first of about 20 successful space missions to Mars were Mariner 4 (which flew past in 1965) and Mariner 9 (which entered Mars orbit in 1971). Mariner 9 recorded massive chasms and volcanoes, including Olympus Mons, the largest volcano in the Solar System. Viking 1 and Viking 2 were the first spacecraft to land successfully on Mars (1976), showing a bleak, rocky, desert landscape. Mars Pathfinder was the first successful mission (1997) that released a rover on Mars that could seek out interesting features near the landing site. Our modern knowledge of Mars comes from the rovers Spirit, Opportunity and Phoenix and the orbiters Mars Express, Mars Odyssey and Mars Reconnaissance Orbiter.

The picture of Mars that emerges from this immense effort is of a planet which now is desert-like, dusty and desiccated at its surface, except at the poles where there is ice, but showing some small signs of water activity such as

Below *Olympus Mons is the biggest volcano on Mars, indeed in the Solar System. Its base is 600 km across and it rises 26 km above the planet's surface, three times as high as Earth's biggest volcano Mauna Loa in Hawaii.*

Right *Mars's north polar cap was imaged by the Mars Global Surveyor spacecraft in 2002, in the Martian spring, when high winds kick up dust storms (bottom centre of the image). The white polar cap is frozen carbon dioxide.*

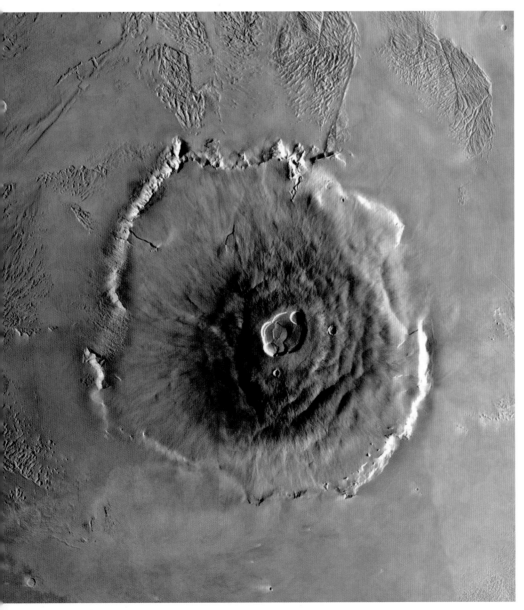

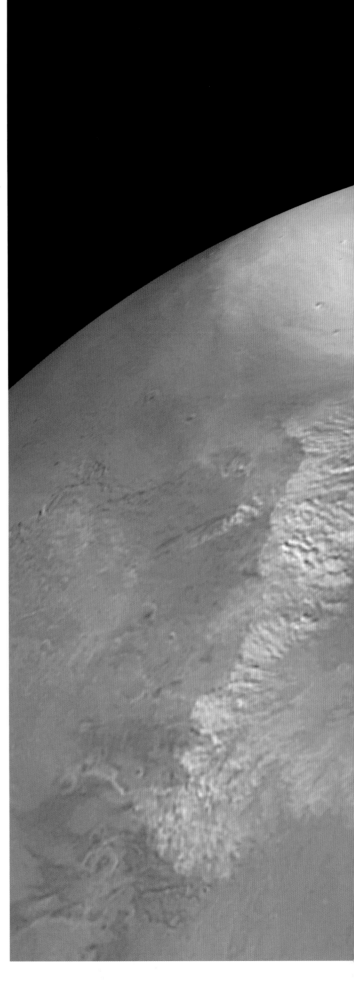

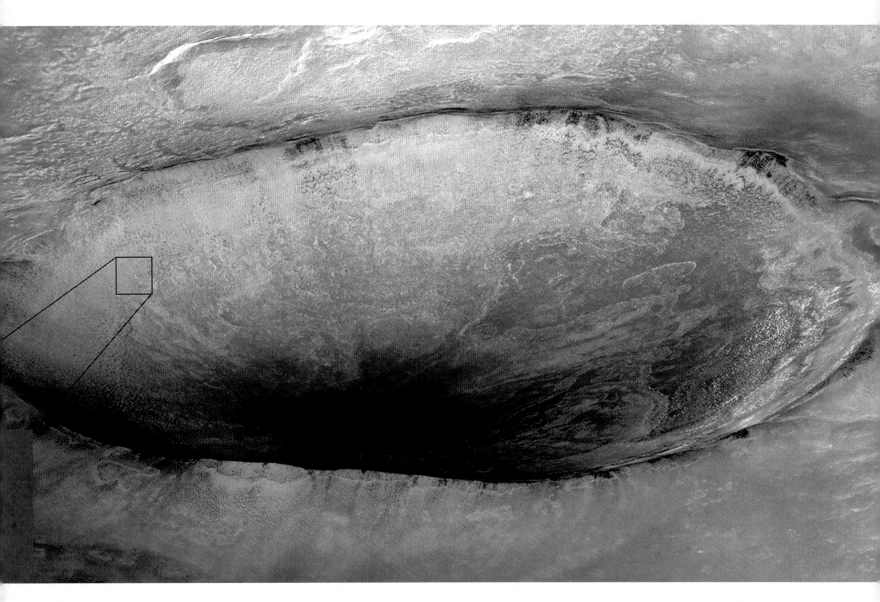

seepage from underground springs that are exposed on the walls of meteor craters on the Martian surface, as well as minerals that form only with water. But there are numerous indications that Mars used to have water in abundance – there are the geological remnants of lakes, glaciers, water running in channels with tumbled, rounded rocks on the stream beds, and massive floods.

The wet epoch of Mars' history is called the Noachian era, ending 3.6 billion years ago. It is named after Noachis Terra (Noah's Land), the region of Mars that shows the most extensive geological features that reveal the action of water, itself named after the Biblical survivor of the Flood.

Is there water on Mars now? There are quantities of ice below the Martian surface, as exposed by the Phoenix rover in 2008 as it dug into the soil. Some meteor craters, like the one known as Yuty, are surrounded by ramparts that look as if they were formed from frozen soil briefly melted by the impact.

The dusty Martian atmosphere is thin and mainly carbon dioxide and nitrogen. Intriguingly, however, as discovered by the Mars Express Orbiter in 2004 there are also traces of methane. On Earth this gas is emitted by active volcanoes (there are none known on Mars) and by biological activity (the gas's common name is "marsh gas", because it is emitted by rotting vegetation, but it is also emitted by belching and flatulent animals). Because the gas is so unstable, it does not last long even in Mars' thin atmosphere, so something on Mars – life? – is making methane even now.

Above *The High Resolution HiRISE camera on board Mars Reconnaissance Orbiter acquired this image of the rover, Phoenix, descending by parachute to the Martian surface. The meteor crater 20 km (12.4 miles) in the background is Heimdall, 10 km (6.2 miles) in diameter.*

Above *The Hubble Space Telescope viewed Mars at its closest approach to Earth on 27 August 2003, summer in the Southern hemisphere of Mars, when its Southern polar cap of white carbon dioxide and water ice has shrunk to its minimum size. The Northern Hemisphere is covered with fresh, uniform volcanic dust. By contrast the Southern Hemisphere is still pockmarked with ancient craters made by meteor impacts, which appear dark because they are filled with coarse sand.*

THE DRY PLANET

What happened to Mars to make it so dry after such a wet past? A surprise from measurements made by the first Mars probes was that Mars does not have any extensive magnetic field, probably because its iron core has frozen solid – the Earth's magnetic field is produced by the motion of its extra-large liquid iron core. The magnetic field of a planet defends its atmosphere from the action of solar particles that emanate from the Sun. Earth's magnetic field deflects these particles (so the large core that gives us deadly earthquakes also gives us life-sustaining air). But Mars' magnetic field became so weak that the solar particles stripped away its atmosphere, evaporating its water. Mars is indeed an almost dead world, having suffered climate change almost beyond any imagining.

The gas giants

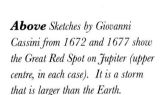

Jupiter

Jupiter is the largest planet, over twice the mass of all the other planets together. Its density is low and it is composed mostly of light gases. What seems like the planet's surface is actually the top layer of clouds in an atmosphere that extends deep within the planet – there is no surface as such, just levels of progressively increasing density. The cloud tops are striped with knotted bands, which lie parallel to its equator. One large area is called the Great Red Spot, first glimpsed by the English polymath Robert Hooke (1635–1703) in 1664 and studied by Giovanni Domenico Cassini (1625–1712) at the Paris Observatory. The Great Red Spot is a storm in Jupiter's atmosphere that has persisted for at least 350 years.

Observed in a telescope, Jupiter is noticeably flattened at the poles, because it rotates rapidly with a period of just under 10 hours. There is a solid core to the planet, to which Jupiter's magnetic field is linked. This magnetic field produces radio static, discovered in 1955 by radio astronomers Bernard Burke (1929–) and Kenneth Franklin (1920–2007) of the Carnegie Institution in Washington DC.

Jupiter's magnetosphere was first directly observed during a flyby in December 1973 by the Pioneer 10 spacecraft and, as is the case with all Jupiter's properties, was studied in detail by the Galileo spacecraft, the only one to stay in orbit around Jupiter for an extended period (from 1995 to 2003). Jupiter's magnetic field extends far into space, almost to Saturn. One of Jupiter's satellites, Io, emits volcanic material which is channelled by the magnetic field on to the top of the planet's atmosphere near the poles. This produces a Jovian version of the Northern (and Southern) Lights.

Jupiter's massive size makes it a big influence on the orbits of comets. In July 1994, it captured, broke up and then swallowed Comet Shoemaker-Levy 9. As the 20 pieces of the comet plunged, one after another, into Jupiter's atmosphere, they bored temporary holes. Atmospheric gases rushed back into the holes, spewed Jovian material from below up through space and over the upper surface of the Jovian clouds, and stained them with transient spots of reddish-brown ices. Having seen what happened in this case, astronomers have identified similar events where transient stains have been seen, but no comet was noticed; one such event was discovered in 2009, by Australian amateur astronomer Anthony Wesley.

Above *Sketches by Giovanni Cassini from 1672 and 1677 show the Great Red Spot on Jupiter (upper centre, in each case). It is a storm that is larger than the Earth.*

Right *The Great Red Spot in a false colour image from the Galileo spacecraft that registers methane in Jupiter's atmosphere in a way that indicates cloud height. The highest clouds are white, pink indicates a higher haze than blue, and black the deepest recesses into the Jovian atmosphere, like the collar around the Great Red Spot.*

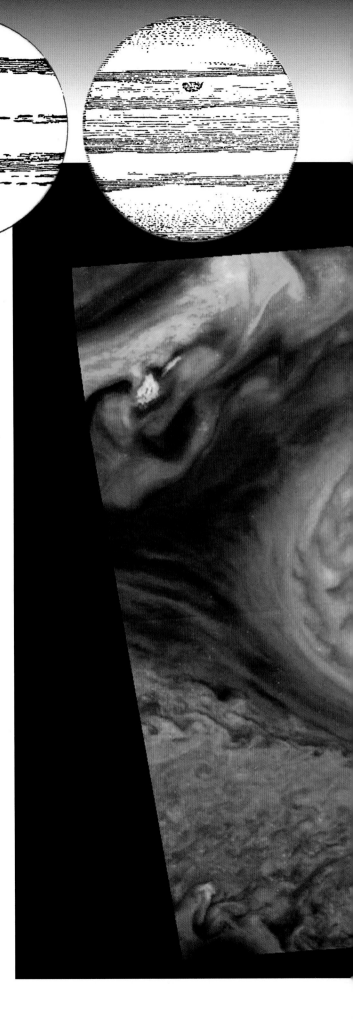

Jupiter is three-quarters hydrogen and one quarter helium; just one per cent of its mass is made up of the heavier elements. It is easy to see why it, like Saturn, Uranus and Neptune, is called a gas giant. The reason why these planets have such a preponderance of light gases is that they lie outside the vaporization zone within which the Sun evaporated light gases and melted ices at the time of planetary formation.

Jupiter's largest satellites, discovered by Galileo in 1610, form a family of four similar-sized worlds, the largest of about 60 moons in Jupiter's family. The two Voyager spacecraft which visited Jupiter in 1979 saw that the outer two (Callisto and Ganymede) are rocky bodies, held together with ice, and covered with impact craters. The second satellite outwards from Jupiter is Europa, which has the distinction of being one of the most spherical objects in the Universe – it is also basically a rocky, icy body, but some of its ice has melted, risen to the surface and formed an ice-covered ocean, which contains more water than the Earth's oceans.

Europa's icy surface has cracked and water has seeped up through the cracks, staining the surface with coloured salts. There are few meteor craters on Europa's surface – when a meteor does make a crater in the ice, the shifting ice floes quickly erase the traces.

The innermost of the Galilean satellites, Io, is startlingly different from its three siblings. A young Voyager navigation engineer, Linda Morabito (1953–), discovered in 1979 that it has active volcanoes that eject ash high above the satellite's surface. Io's surface is covered with volcanic craters, lava flows and drifting ash.

The reason why Europa and Io are so different from the cold, rocky satellites Callisto and Ganymede is that they are close to Jupiter and experience its strong tidal force. They are compressed and released, pumping like a heartbeat. The energy of the movement releases heat, and raises the satellites' internal temperature.

Below *Comet Shoemaker-Levy 9 was ripped into several pieces by the tidal force of Jupiter, each of which crashed in sequence into its atmosphere in 1994.*

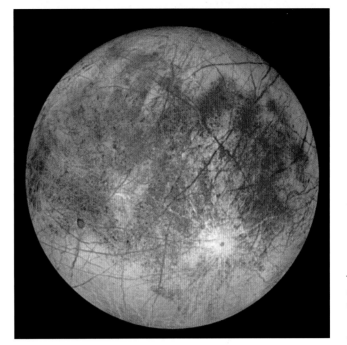

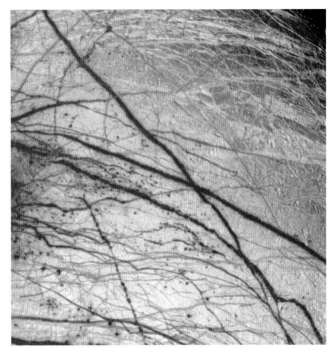

Opposite *Banded clouds form a complex wake downstream of the vortex of the Great Red Spot in this picture taken by the Cassini spacecraft in 2000. The clouds are made of ammonia, hydrogen sulphide and water, with different compounds and elements brought up from below, giving beautiful brown, pink and orange colours.*

Above left *The surface of Jupiter's satellite Europa is ice-covered ocean, criss-crossed by rifts and cracks in the ice floes, where upswelling water has deposited coloured salts. A meteor has recently struck lower right and scattered fresh snow on top of the ice.*

Left *On Europa, irregular ice floes shift and grind against one another to slosh water onto the top of the ice and create coloured stains.*

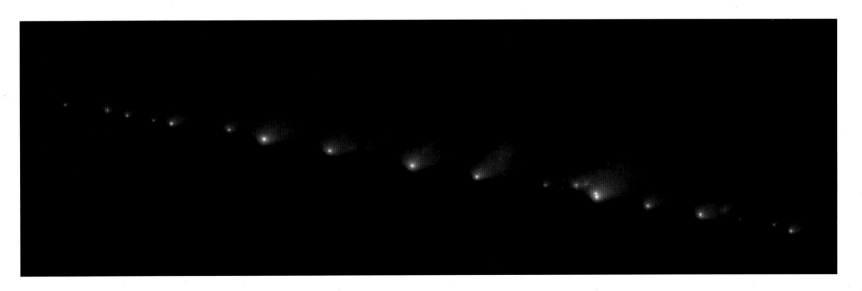

Saturn

When Galileo saw Saturn in his telescope in 1610, he could see something strange either side of the planet. He thought that maybe it had two large moons. Over the decades the appearance of the planet changed, and the "moons" disappeared. We now know that Saturn is surrounded by thin rings, which all but disappear when the Earth passes across their plane. It was in 1659, that the Dutch physicist Christiaan Huygens, with his greatly improved telescope, was able to see the form of the rings. He also discovered the largest satellite of Saturn, Titan.

Saturn was first explored by spacecraft in brief flybys by Pioneer 11 in 1979 and by the two Voyager spacecraft in 1980–81. These spacecraft established that Saturn had Jupiter-like clouds and weather, though Saturn's greater distance from the warmth of the Sun means they are not so active. They also saw that Titan has an opaque atmosphere. The NASA/ESA spacecraft Cassini arrived at Saturn in 2004 and is expected to stay in orbit there until 2017. On arrival, it released the Huygens Lander to parachute on to Titan and establish the nature of its atmosphere and surface.

Saturn has more than 50 recognized satellites, as well as the almost countless number that make up the rings. Of these, Mimas is remarkable in having a gigantic crater caused by the impact of an asteroid of such size that it must have come close to destroying the moon – indeed the shock of the impact created grooves and chasms all over Mimas's surface. Another of Saturn's satellites, Enceladus, has geysers that shoot water and icy material well above its surface into space.

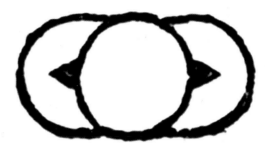

Opposite *The south pole of Io shows several volcanic calderas, lava flows and drifts of ejected white sulphur dioxide snow.*

Left *Early telescopes did not have fine enough resolution to identify exactly what was unusual about Saturn. This image was published in 1622 by the Italian physician Fortunio Liceti.*

THE RINGS OF SATURN

For centuries after their discovery, the nature of Saturn's rings remained unclear. In 1787, the French mathematician Pierre-Simon Laplace (1749–1827) suggested that the rings were made of solid ringlets, but in 1859, in response to a challenging question posed for a prize, the Scottish mathematician James Clerk Maxwell (1831–79) showed that solid rings would break up and fall apart. He suggested that the rings are made up of numerous particles, each a small moon orbiting the planet. His theory was shown to be correct in 1895, when astronomer James Keeler (1857–1900) of Allegheny Observatory found that the inner parts of each ring were moving faster than the outer parts, as expected if the rings are made of innumerable small satellites. The most widely held theory of their origin, proposed by French astronomer Édouard Roche (1820–83), is that they are fragments of one (or more) moons of Saturn which came too close to the planet and were destroyed by tidal forces. The Cassini spacecraft has established the role of some of Saturn's satellites in creating the pattern of gaps in Saturn's rings, and braiding some of the thin rings into twists through their shepherding action on the herds of individual particles of which the rings are composed. It is remarkable that more than two centuries after the force of gravitation was discovered scientists are still discovering some of its subtle effects.

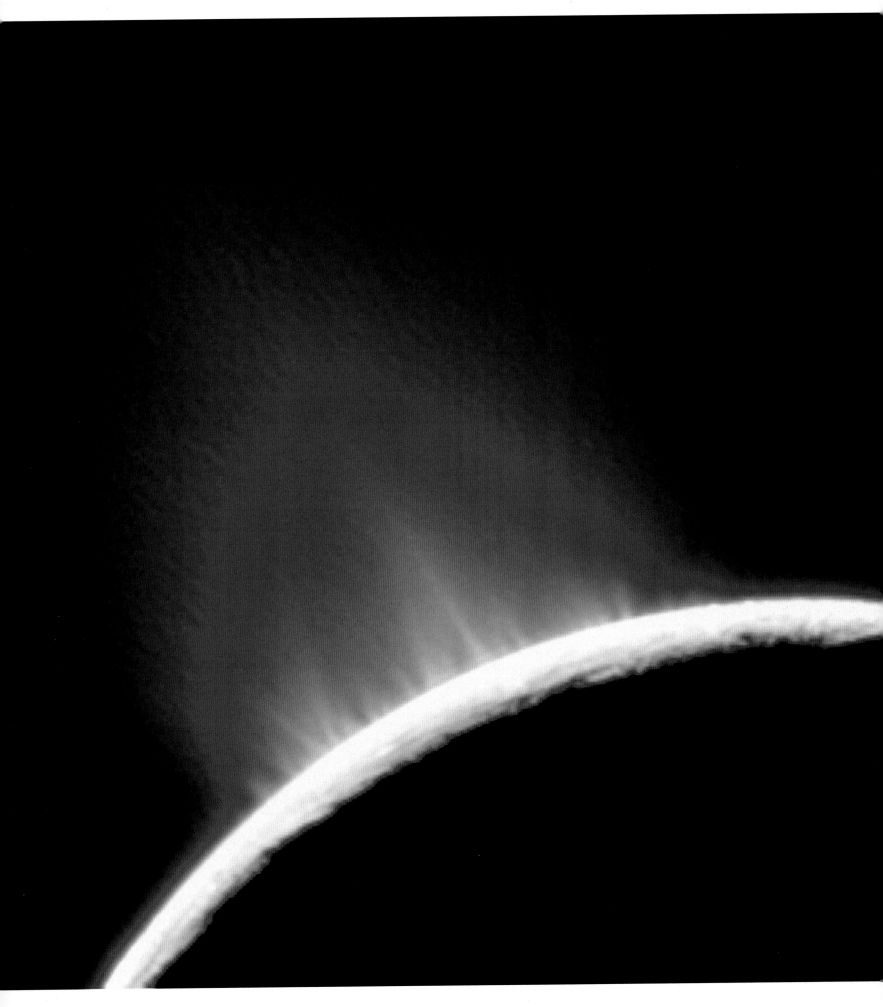

Saturn's main moon, Titan, is a rocky, icy world, whose polar regions, revealed by the Huygens probe in 2004, are covered with liquid hydrocarbon lakes. Its atmosphere is mainly smog-filled nitrogen, with methane and ethane clouds. It is windy and it rains liquid hydrocarbons, forming rivers, and lakes, with associated geological features like lake shores, sand dunes and river beds. The atmosphere of Titan is like the Earth's atmosphere used to be, and the rich carbon chemistry of its atmosphere and lakes is thought to resemble that which prefigured life on Earth.

Uranus and Neptune

The modern space exploration of Uranus and Neptune is at an early stage; they have each been visited only once in a flyby by the Voyager 2 spacecraft in 1986 and 1989 respectively. Uranus rotates about an axis that lies in the plane of its orbit and its seasons are both extremely variable and bizarre; it is possible that the planet was knocked over by an asteroid impact. Uranus has a rudimentary ring system like Saturn and a retinue of moons, including Miranda, which has chaotic terrain that suggests that it too has suffered a catastrophic impact. The impact broke up the satellite, but only just, so that it reassembled in a lopsided sort of way. Neptune has surprisingly active weather, even though it is so far from the Sun in the cold distant regions of the Solar System, and its moon Triton has cryovolcanoes (cold volcanoes) that erupt geysers of water, nitrogen, ammonia and methane.

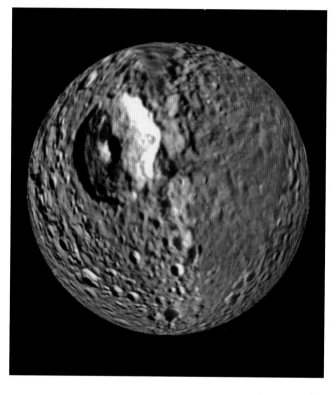

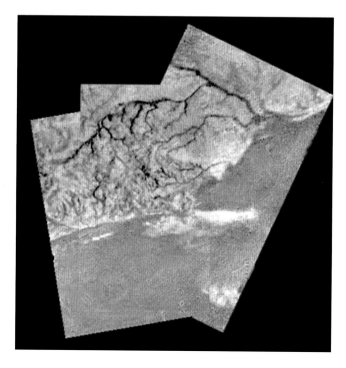

Opposite *Ice geysers erupt on Enceladus, back lit by the Sun.*

Above left *Saturn's moon Mimas is dominated by a single huge crater, Herschel, caused by a meteor or asteroid that must have nearly disrupted the satellite.*

Far left *Titan is Saturn's large icy moon. In this view by ESA's lander Huygens, liquid methane rivers have carved their tributaries through a ridge area into a larger channel or lake.*

Left *Uranus' moon Miranda is scarred and twisted. Its patterns of ridges and bands enclose less disturbed areas, and the whole world seems to have been reassembled wrong after being shattered by a giant impact, or, alternatively, mangled by fierce tidal forces in an earlier orbit.*

Other planetary systems

Copernicus theorized in 1543 that the Sun was a star and the Earth a planet. The first person explicitly to suggest that there might be planets in orbit around other stars was the Italian Dominican friar and philosopher, Giordano Bruno (1548–1600). The then Archbishop of Canterbury derided him for supporting "the opinion of Copernicus that the Earth did go round, and the heavens did stand still; whereas in truth it was his own head which rather did run round, and his brains did not stand still". Bruno was tried by the Inquisition for his heretical theology and burnt at the stake in Rome in 1600, his tongue imprisoned because of his wicked words. But Bruno's astronomical ideas were echoed by Isaac Newton (1643–1727) in his *Principia* (1713 edition): "And if the fixed stars are the centres of similar systems, they will all be constructed according to a similar design and subject to the dominion of One."

The first clear discovery of planets in orbit around another star was the discovery in 1992 by the Polish radio astronomer Aleksander Wolszczan (1946–) and his colleague Dale Frail

__Below__ The star Fomalhaut has been blocked out at the centre of this HST picture, and is orbited by a circular disc of dust (seen obliquely and looking elliptical). A tiny dot on the edge of the dusty disc (in the box) is a planet 17 billion kilometres from its star. As shown in the insert, the planet moved in its orbit between 2004 and 2006; it takes 872 years to complete one revolution.

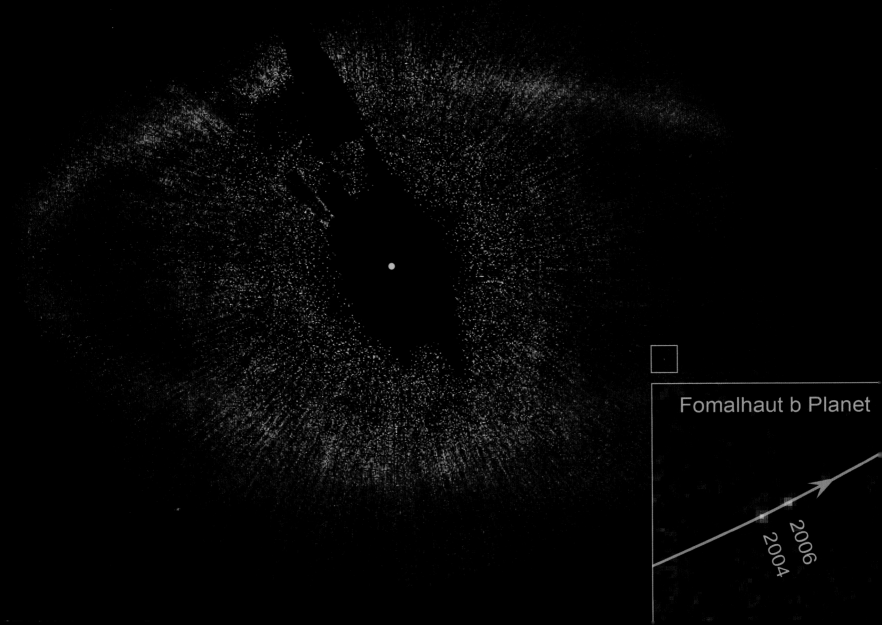

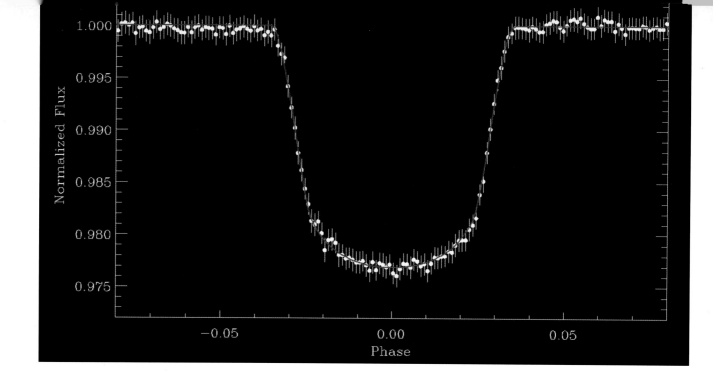

Above *The first planet discovered by the CoRoT satellite showed itself through the "wink" (graphed here) of a star, which dimmed by just 2 per cent as its planet transited in front. The planet, which is 1.8 times the diameter of Jupiter, orbits its star in 1.5 days, so it must be very close to it and very hot.*

Right top and bottom
"Before" and "during" pictures of a star field showing one star brightening as another passes nearly exactly in front. The nearer star, and any planets that it has, act as a succession of gravitational lenses, magnifying the image of the star behind in a series of pulses.

working with the Arecibo radio telescope in Puerto Rico. The planets are in orbit around a pulsar, PSR 1257+12. The three planets may be survivors of the supernova explosion that created the pulsar but were probably formed from the debris, and thus are a planetary system formed at the death of a star, not at its birth, as was our Solar System.

In 1995, Swiss astronomers Michel Mayor and Didier Queloz discovered the first planetary system like the Solar System, using a modest telescope of the Observatoire de Haute-Provence in France. They had developed an exquisitely accurate instrument to detect the slight wobble of the parent star from one side to the other, which is caused by the gravitational pull of its daughter planet in orbit.

Further extrasolar planets revealed themselves through the wink of a star as its daughter planet transits in front, blocking out a minute fraction of its light. Ground-based telescopes find this difficult, because small changes in the atmosphere confuse the measurements, but two satellites, the mainly French CoRoT telescope (launched 2006) and the NASA Kepler mission (launched 2009) have had good success working in the clarity of space.

Some extrasolar planets have been discovered by watching the astonishing brightening of a star by the transit of another star that lies at random along the line of sight as it passes in front. The background star brightens because the nearer star makes a gravitational lens that magnifies the image of the star behind. If the nearer star has a retinue of planets and the alignments with them are within a whisker of perfect, the background star brightens in bursts.

Most of the 500 extrasolar planets known are Jupiter-sized planets, orbiting surprisingly close to their parent star, roasted and evaporating. This statistical fact is heavily influenced by the fact that astronomers can only detect the bigger, more massive planets in short-period orbits, because the wobbles and winks produced by such stars are larger and more frequent.

Our Galaxy and others

With his telescope, Galileo saw that the Milky Way was composed of stars too faint and crowded to be perceived individually. But why are faint stars concentrated into one band? The first person successfully to answer this question was Thomas Wright (1711–86), an English astronomer. In his book *An Original Theory or New Hypothesis of the Universe* (1750), he described the Milky Way as "an optical effect due to our immersion in what locally approximates to a flat layer of stars". Having read a newspaper report of Wright's theory, philosopher Immanuel Kant (1724–1804) gave the idea further currency in his *Universal Natural History and Theory of Heaven*.

In 1784, astronomer William Herschel mapped out the precise shape of the flat slab-like distribution of stars by counting them in different directions across the Milky Way. The Sun was only slightly off-centre. Herschel did not know that dust lies between the stars, restricting the view: in a fog, it always appears that you are in the middle of what you see.

Herschel's work was developed by the Dutch astronomer Jacobus Kapteyn (1851–1922) using essentially the same techniques. He concluded that the stars were distributed in a flat, frisbee-like shape, some 60,000 light years in diameter and 10,000 light years thick – the grand size of the Milky Way system of stars ('the Galaxy', for short) had started to become apparent.

The American astronomer Harlow Shapley (1885–1972) brought a fresh approach to the topic. He was studying globular clusters of stars and noticed that they lay above and below the Milky Way, but most appeared in one direction, towards the constellation Sagittarius. From this, in 1921 he proposed that globular clusters are distributed around a central point of the Galaxy (in the direction of Sagittarius), with the Sun off-centre. He had developed a way to estimate the distances of globular clusters, based on some variable stars often found in them, and estimated that the Milky Way was flat in shape and 300,000 light years in diameter. This is more or less the modern picture of the Galaxy.

Below *In 1785 William Herschel counted the density of stars in various directions in a section across the Milky Way, showing that Wright's picture was basically right, but also that the plane of the Milky Way was cleft in two (we now know this to be due to dust) and that the Sun (larger dot, near centre) was off centre.*

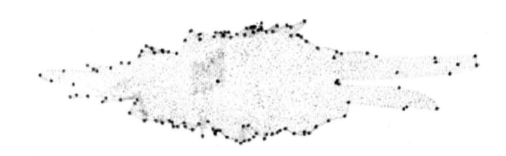

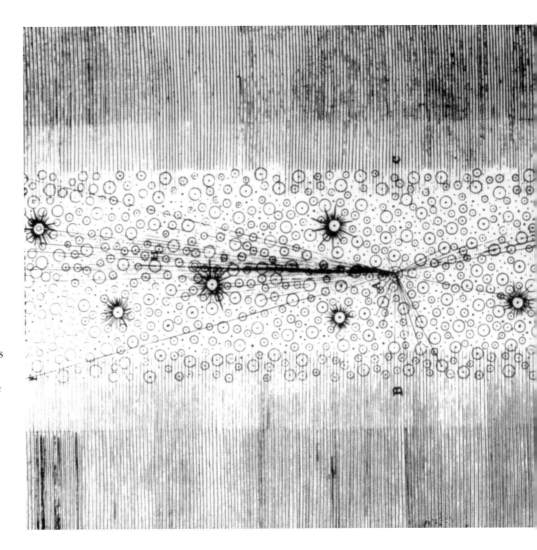

Right *In 1750 Thomas Wright envisaged the Milky Way as a flat disc-like slab of stars, with sight lines lying within the slab viewing a large number of distant stars and producing the "milky" effect.*

Right *Thomas Wright suggested that our Milky Way was one of a number of independent star systems, somewhat like galaxies, although his imagination did not grasp the vast volume of empty space that actually lies between them.*

Overleaf *The Milky Way arches above the European Southern Observatory, split along its length by dark dust clouds. Shining up from the rising Moon is the zodiacal light (also dust, lying in the plane in which the planets orbit, reflecting sunlight), and to the right, below the arc of the Milky Way, are two of our galactic neighbours, the Large and Small Magellanic Clouds.*

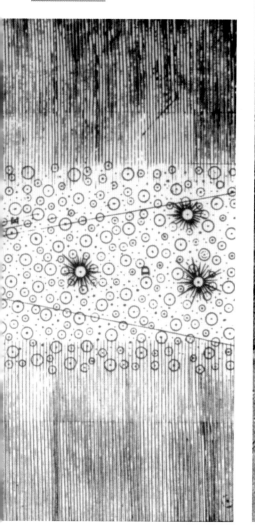

Above *The Andromeda Galaxy, M31, is a spiral galaxy similar to our own Milky Way, seen at an oblique angle. This image has been taken in ultraviolet light with the Swift satellite, emphasising the young hot stars in the spiral arms.*

Thomas Wright had related his picture of the Milky Way to the discovery of the many faint nebulae that had been discovered around the sky. He described "the many cloudy spots, just perceivable by us, as far without our Starry regions, in which tho' visibly luminous spaces, no one star or particular constituent body can possibly be distinguished; those in all likelihood may be external creation, bordering upon the known one, too remote for even our telescopes to reach". In other words, the nebulae are not cloudy masses, at all, they are vast but distant collections of individual stars like our own Galaxy. In 1755, Immanuel Kant introduced the term "island universe" for this concept.

One such nebula is known as M31, the Great Nebula in Andromeda. In 1885, an unusual nova was observed in M31, with further novae observed by George Ritchey (1864–1945), and in 1917 that the American astronomer Heber Curtis (1872–1942) estimated the distance of these stars compared with similar stars in our Milky Way. He came to the conclusion that M31 must be some 100 times further than distant stars of the Milky Way. Looking back through old photographs of M31, he found other examples, much fainter than typical examples of such stars in our own Galaxy. He calculated that M31 was 100 times further away than a typical star in our Galaxy, in line with the concept that M31 was an "island universe". In 1920, Shapley and Curtis took part in what became known as the Great Debate at the Smithsonian Museum in Washington, DC, on the nature of the Milky Way and the nebulae. Curtis is usually reckoned to have had the best of the debate, although it

RADIO WAVES FROM THE GALAXY

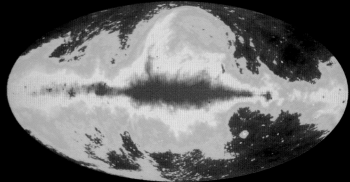

During the Second World War (1939–45), the Netherlands was occupied by an enemy army, with suspicions, shortages, curfews, bans on gatherings and other inconveniences (and much worse). It was not possible for Dutch astronomers to work with telescopes at night. Astrophysicist Jan Oort (1900–92) organized discreet meetings of the Dutch Astronomy Club to discuss theoretical questions that could be tackled at home by means of brains, pencil and paper. At one meeting, he set his student Hendrik van de Hulst (1918–2000) the problem of whether hydrogen, which Oort knew to be common throughout the Galaxy, emitted radio waves with detectable intensity.

Hulst found that hydrogen atoms emit radio waves at 21-cm (8-inch) wavelength. The emission happened very rarely in any given atom – once every 10 million years in a lone hydrogen atom floating in space – but this rarity was compensated by the vast number of hydrogen atoms. After the war had ended, 21-cm hydrogen emission was detected in 1951 simultaneously by United States radio astronomers Harold Ewen (1922–) and Edward Purcell (1912–97) at Harvard University, Dutch astronomers Alexander (Lex) Muller (1923–2004) and Oort, and Australian radio astronomers Wilbur ("Chris") Christiansen (1913–2007) and Jim Hindman.

was somewhat inconclusive with each astronomer stumbling at times. The debate identified key issues and led to the work later by Edwin Hubble (1889–1953), using the new 100-inch Mount Wilson telescope. He was able to see individual stars in some of the nearer nebulae, including some Cepheid variable stars. Assuming they were similar to the variable stars in the globular clusters in our Galaxy, he determined the distances of the nebulae. The "island universes" were definitely outside the Milky Way.

It is difficult for to see the whole Galaxy through the fog of interstellar dust, but radio waves have revealed it in its entirety. During the 1950s, radio astronomers were able to map the entire Galaxy through the radio waves emitted by interstellar hydrogen. They reveal that the Milky Way Galaxy is a spiral galaxy, with our Sun 25,000 light years from its centre.

Above *A radio map of the whole sky, colour-coded with red areas being brighter than blue. The plane of the Galaxy runs left to right along the centre line of the map, and the arc reaching up to the top is known as the North Polar Spur, the remains of a long-ago nearby supernova.*

The universe of galaxies

A small number of galaxies can be seen with the unaided eye. The earliest recorded observation of one was by the Islamic astronomer Abd al-Rahman al-Sufi (903–986) in 964, who described a "small cloud" in Andromeda in his *Book of Fixed Stars*. It is a galaxy much like our Milky Way galaxy. The two nearest galaxies to us are satellites of the Milky Way galaxy, called the Magellanic Clouds.

These three galaxies and our own are members of a group of more than 40 galaxies that comprise the Local Group, lying within a distance of about five million light-years from Earth. The Local Group is part of and lies on the edge of a much richer cluster of up to 2,000 galaxies whose centre lies in the direction of the constellation Virgo at a distance of some 60 million light-years. Thirteen of the brightest galaxies are mentioned in a catalogue of just over 100 nebulae compiled by French astronomer Charles Messier (1730–1817) and published in several editions between 1774 and 1784. A very large catalogue of nearly 8,000 relatively near-by nebulae – mostly galaxies – was compiled by the Irish astronomer John Louis Dreyer (1852–1926) in 1888, principally from the all-sky surveys made by William Herschel (1738–1822) and his son John Herschel (1792–1871).

Right *The grand-design spiral galaxy, M51, known as the Whirlpool Galaxy, with a smaller companion galaxy, NGC 5195, passing under the tip of one spiral arm.*

Opposite insert *The sketch made in April 1845 with William Parsons's telescope within a few weeks of its first light that established the name of the Whirlpool Galaxy and the spiral nature of some galaxies. The sketch stands comparison with the Hubble Space Telescope picture made 150 years later.*

Left *William Parsons, the Third Earl of Rosse, creator of what was in 1845 the largest telescope in the world (see page 66), with which he and his team discovered the spiral nature of some galaxies.*

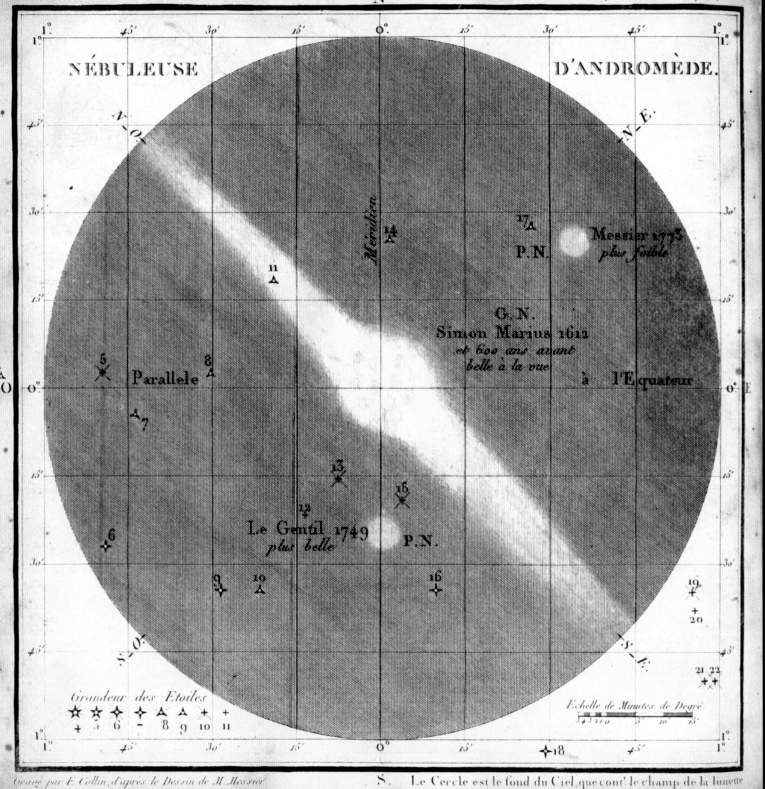

Above Charles Messier charted the Andromeda Nebula, now known to be a galaxy, with two companion galaxies (bottom left and top right). The three objects are M31, M32 and M110 (numbers 31, 32 and 110 in Messier's catalogue). Messier has added descriptions of the three galaxies, including their discoverers, thus documenting his own discovery of M110 in 1773. The discovery of the Andromeda Nebula itself he attributed to Simon Marius, who gave a detailed description of it in 1612.

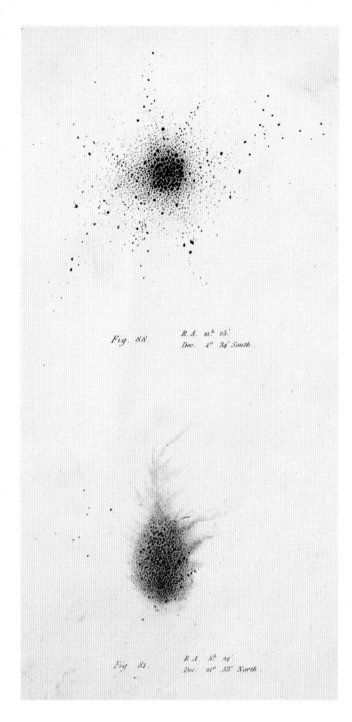

These nearer, brighter galaxies were the ones that were the subject matter of studies of galaxies for two centuries and formed the basis of our picture of the content of the large-scale universe. At first perceived as amorphous fuzzy blobs, galaxies started to take shape when viewed with large enough telescopes, starting with the so-called Leviathan of Parsonstown, a telescope with a six-foot (1.8-metre) mirror erected by William Parsons (1800–67, the 3rd Earl of Rosse) at Birr Castle in Ireland. It was a project of what would now be called Keynesian economics, intended to alleviate the poverty of a country estate depressed by the Irish potato famine. It created what became the largest telescope in the world for half a century.

Parsons and his assistants were able to make out spiral arms in some galaxies, starting in 1845 with the so-called Whirlpool Galaxy. As more and more examples of spiral galaxies were identified, it became clear that they were flat discs seen at random angles with a generally elliptical profile. The spiral pattern suggested that the discs were rotating. The

Above *The Irish astronomer William Parsons, the Third Earl of Rosse, drew a sketch of the globular cluster of stars, M2 (top), using his 36 inch (90 cm) reflecting telescope at Birr Castle. He noted that, because of his large telescope, the stars in it were more numerous and brighter than had ever been seen before. In the same plate, he contrasted M2 with the Crab Nebula (bottom), which had been thought to be a similar star cluster at such a large distance so that its individual stars could not be discerned, but, it seems from the representation, thought he was on the verge of doing so. In fact, the nebula is not made of stars but of gas. The nebula got its name from its appearance in this drawing, with claw-like filaments extending out from the southern extremity (towards the top). Later, the drawing's creator repudiated his own representation of the nebula as a crustacean – "I would have figured it different", he ruefully wrote after seeing it more clearly through his larger six-foot (1.8 metre) reflector. But by then the name had stuck.*

Below *M51, the Whirlpool Galaxy, is a wonderful grand-design spiral galaxy 30 million light years away, with this composite image constructed from NASA's Chandra X-ray Observatory (X-radiation showing as a purple colour), GALEX (ultraviolet radiation showing blue) the Hubble Space Telescope (optical as green) and the Spitzer Space Telescope (infrared as red). The purple X-ray sources are black holes and neutron stars in binary star systems. The blue highlights clusters of hot, young stars. The red nebulae are clouds of dust and gas, associated with some bright optical nebulae in green. M51 is interacting with a close neighbour galaxy, NGC 5195, at top, which is drawing one of M51's spiral arms away from its spiral track.*

invention of photography and its application to astronomy by pioneers such as Welsh-born engineer and amateur astronomer Isaac Roberts (1829–1904) and United States doctor and amateur astronomer Henry Draper (1837–82) meant that greater details could be seen in the galaxies, since the images could be examined at leisure and measured. For example, Roberts's photograph in 1884 of the Andromeda Nebula revealed that it is a spiral galaxy seen at an oblique angle. Sometimes, in the more face-on examples of spiral galaxies (where the shape could be more accurately seen) the spiral arms started, not from the centre of the galaxy but from either end of a "bar" across the middle – not unnaturally, these were known as "barred spirals". Naturally, given their beauty, spiral galaxies attracted a lot of attention; those that have a strong degree of symmetry are called "grand design"' galaxies and are often photographed.

At first, attention focused strongly on these galaxies, not only because of their beauty but also because of the so-called "nebular hypothesis" of the Swedish scientist Emanuel Swedenborg (1688–1772), the Prussian philosopher Immanuel Kant (1724–1804) and the French mathematician Pierre-Simon Laplace (1749–1827) of the origin of the Sun and the Solar System – the suspicion was that the "spiral nebulae" might be exemplars of this hypothesis. When the distances (and therefore the sizes) of the "spiral nebulae" became clear, this was ruled out – they were the size of entire stellar systems rather than of stars and their planetary systems. It turned out, moreover, that there was another, more numerous population of (usually larger) galaxies which showed no trace of spiral arms. They, too, had an elliptical shape, but that was, it was determined, because they were three-dimensional ellipsoids (prolate ellipsoids like American or Rugby footballs, oblate ellipsoids like tangerines, or triaxial ellipsoids that were a combination of the two shapes). Seen at random angles these ellipsoids all looked elliptical, some pointier than others which were more circular-looking.

It was United States astronomer Edwin Hubble (1889–1953) who in 1926 brought some sort of order to this confusion of shapes, identifying subclasses of the spiral and elliptical galaxies. Galaxies that did not fit anywhere else in the scheme were placed in a miscellaneous category of "irregular". Often in these cases, two galaxies are passing so

Above *The Hubble Ultra Deep Field (see page 71) shows a motley collection of very distant galaxies, seen as they were a very long time ago. They are smaller, more shapeless than their counterparts today. They have not had time to settle down to their regular form and to absorb other galaxies in acts of cannibalism like those that produced our own Galaxy.*

Opposite *NGC 1132 is a massive galaxy, the amalgamation of a number of smaller galaxies that have merged together. Each contributed a number of globular clusters of stars that swarm around NGC 1132. In the background are many galaxies, much further away.*

close to each other that they distort each other, each pulling stars in sheets and streams from the other.

The brighter galaxies are elliptical or spiral galaxies, but most galaxies are much fainter "dwarf galaxies", a category whose importance has only recently been recognized and whose numbers grow every day as individual examples are discovered. Often dwarf galaxies are in orbit around a larger galaxy, like moons around a planet. The Milky Way galaxy has several satellites. One of them, the Sagittarius dwarf galaxy, is being stretched by its interaction with our Galaxy, its stars plunging in to be absorbed by its larger neighbour.

In extreme cases, colliding galaxies merge together. In fact, astronomers now think that this is the way in which the large elliptical galaxies are formed – spirals that collide lose all their orderliness, their gas gets stirred up and makes stars in a great burst rather than continuously, and, as the stars age and fly everywhere in all directions, an elliptical galaxy of ageing stars results.

It was more than half a century after Hubble's observations before good evidence for this idea about the formation of galaxies became available, through the light-gathering power of the Hubble Space Telescope. One innovation in the way the telescope was used was to give some time to the telescope's director to do with as he chose. In 1995, the then Director of the Space Telescope Science Institute, Robert Williams, chose to get the telescope to stare for 10 days at what until that time was known as an empty area of sky. The HST would then record the faintest possible galaxies, with the idea that "faint means far, means long ago" and that the telescope would be looking back in time as far as it could see. The image is known as the Hubble Deep Field and it had the images of only a dozen stars at the edge of our own Galaxy but recorded 3,000 faint galaxies at distances up to 10 billion light years. Because the light from these galaxies takes so long to cross the vast distance to Earth, we see them only a few billion years after their formation The data and its interpretation proved to be so important that the idea was repeated in 2003 with a more advanced camera and longer exposure, and the resultant Hubble Ultra Deep Field recorded 10,000 galaxies at distances up to 13 billion light years away.

The images contain smaller galaxies than the local region of universe. There were fewer elliptical galaxies than now and the spirals were more disturbed and more irregular. Some of them had not settled into being spirals, others were then packed closer together and had got tangled up with neighbours. The galaxies of the present time have built up over time to their present size from mergers of small, young galaxies like these, settling into grand design spirals and creating ellipticals

Opposite *The Large Magellanic Cloud is a rather anaemic galaxy, but nevertheless its barred spiral shape can be discerned – its bar of stars runs diagonally from bottom left to top right, and emaciated spiral arms, delineated by pink nebulae, run from the tips of the bar, up from the left and down from the right.*

Left *American astronomer Edwin Hubble, dressed for his observing session in clothing to keep him warm against the cold night air, looks through the eyepiece of the 100-inch telescope on Mount Wilson. He is adjusting the telescope's position to keep it tracking perfectly while he exposes a photograph held at the focus of the telescope in the plate-holder above his head.*

Below *Hubble's "Tuning Fork Diagram" is an attempt to make sense of the variety of the shapes of galaxies.*

HUBBLE'S TUNING FORK DIAGRAM

Hubble classified galaxies in what became known as the Tuning Fork Diagram, after the bifurcated scheme in which it was laid out. He separated galaxies into two broad classes – ellipticals and spirals. He classified the ellipticals in a sequence that ranged from very pointy to almost circular (spherical). The spirals and barred spirals had arms that were tightly wound to more loosely wound. For a long time the exact classification of our own Galaxy has not been clear but the present consensus is that it is a large barred-spiral galaxy. Its bar, which we see at a skew angle, is the bright, thick part of the Milky Way in the direction of Sagittarius.

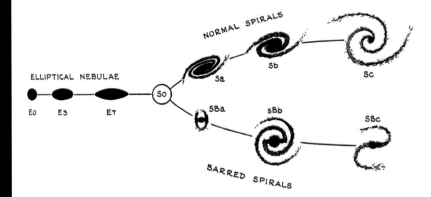

THE MAGELLANIC CLOUDS

Al-Sufi described an object called Al Bakr, the White Ox, adding that while Al Bakr is invisible from northern Arab countries, because it was so close to the south celestial pole, it can be seen from the southern outlet of the Red Sea into the Indian Ocean. This refers to one of two similar structures, one large, one small, which look as if they are pieces of the Milky Way that have broken off. They were first seen by Europeans during the early voyages of discovery to the southern seas. They were drawn with the Southern Cross as "twoo clowdes of reasonable bygnesse" on a star chart in 1516 by an Italian navigator and spy, Andrea Corsali (1487–?), who travelled as a double agent for the Medici family, seeking out commercial possibilities for them on a secret Portuguese voyage to India. Later, the clouds became associated with Ferdinand Magellan (1480–1521), the Portuguese captain who led the first voyage around the world (1519–22). They were noticed by members of his crew and described in accounts of the voyage; unfortunately Magellan himself did not live to tell of them, since he was killed in the Philippines towards the end of the voyage. The two galaxies are now known as the Large and the Small Magellanic Clouds.

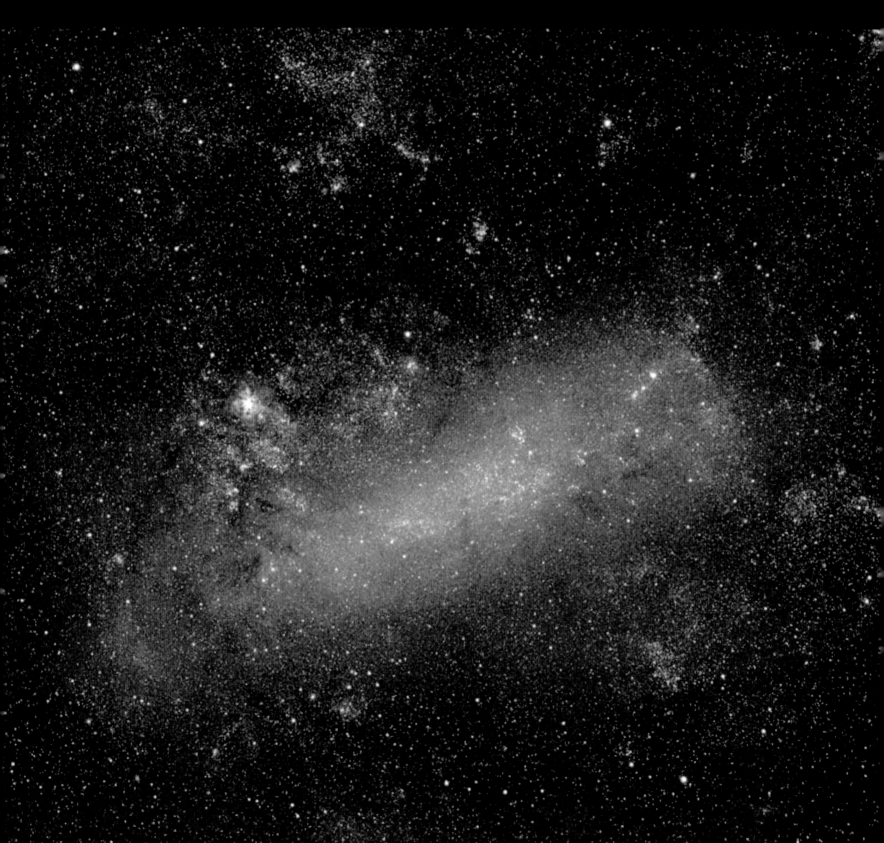

Exploding galaxies and quasars

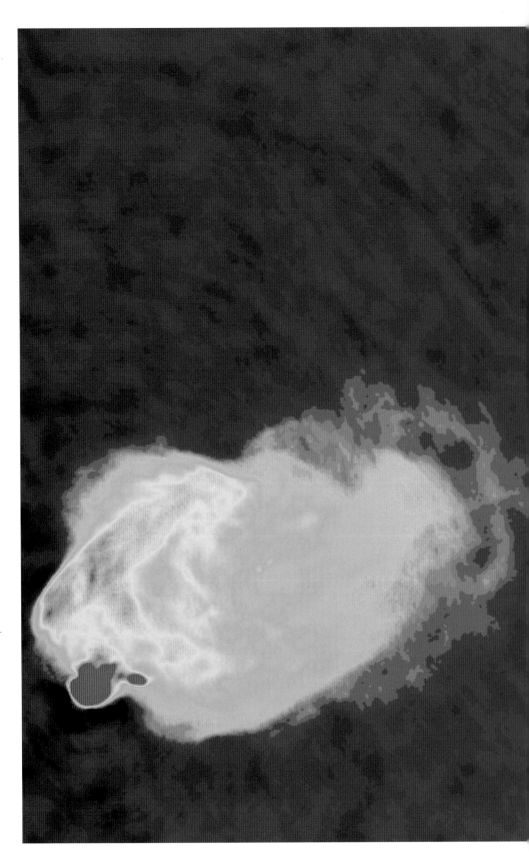

Why do radio-emitting galaxies emit such copious amounts of energy? In 1951, Cambridge radio-astronomer Francis Graham Smith (1923–) accurately measured the position of a radio galaxy named Cygnus A, the strongest in that constellation. United States astronomer Walter Baade (1893–1960) at the California Institute of Technology used the 200-inch Mount Palomar telescope to discover an unusual galaxy at that position. Its shape was a puzzle – there was talk that it was two colliding galaxies, but in reality this shed little light on the problem.

The breakthrough came in 1962. A radio source called 3C273 proved to be a galaxy at a huge distance. A new name for galaxies like it was derived from their technical name, Quasi-stellar Radio Sources – "quasars". Incredibly bright, at incredible distance and incredibly small; the emerging properties of quasars intensified the problem of where their energy originated.

The answer was that the energy comes from a massive black hole central to a galaxy, feeding on gas that is falling in to it. So much energy is released that some gas is bounced around and ejected, and many quasars have jets shooting out at high speeds, like 3C273's wisp, which are stabilized along the axis of a rotating black hole that acts as a heavy, rapidly spinning top – this proved to be the case for Cygnus A when it was examined with modern radio telescopes.

Gas also speeds around a quasar at stupendous speeds. It must be moving close to something that is incredibly massive. A team led by Johns Hopkins University astronomer Holland Ford using the Hubble Space Telescope in 1994 found that the quasar in the galaxy M87 was two to three billion times the mass of the Sun.

There is a recent discovery that actually proves that these massive objects are black holes. Cambridge astronomer Andy Fabian used the Japanese X-ray astronomy satellite Asca in 1995–99 to identify features in the spectrum of the galaxy catalogued MCG-6-30-15 that can only come from effects of relativity in the powerful gravity of a black hole, the strongest evidence yet that black holes really exist.

Right *The exploding galaxy Cygnus A as seen by the Very Large Array radio telescope in a false-colour picture of radio emission in which red is bright and blue is faint. At the centre of the galaxy is a black hole, seen here as the red dot at the middle of the picture, from which two jets of material shoot out and billow up against the surrounding intergalactic gas.*

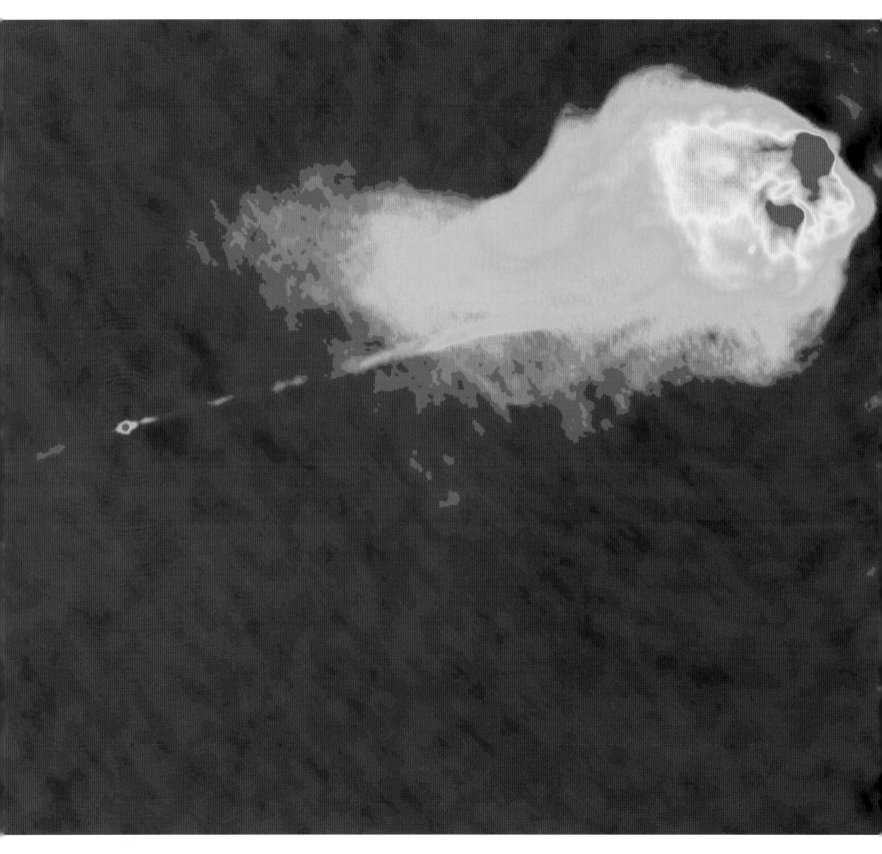

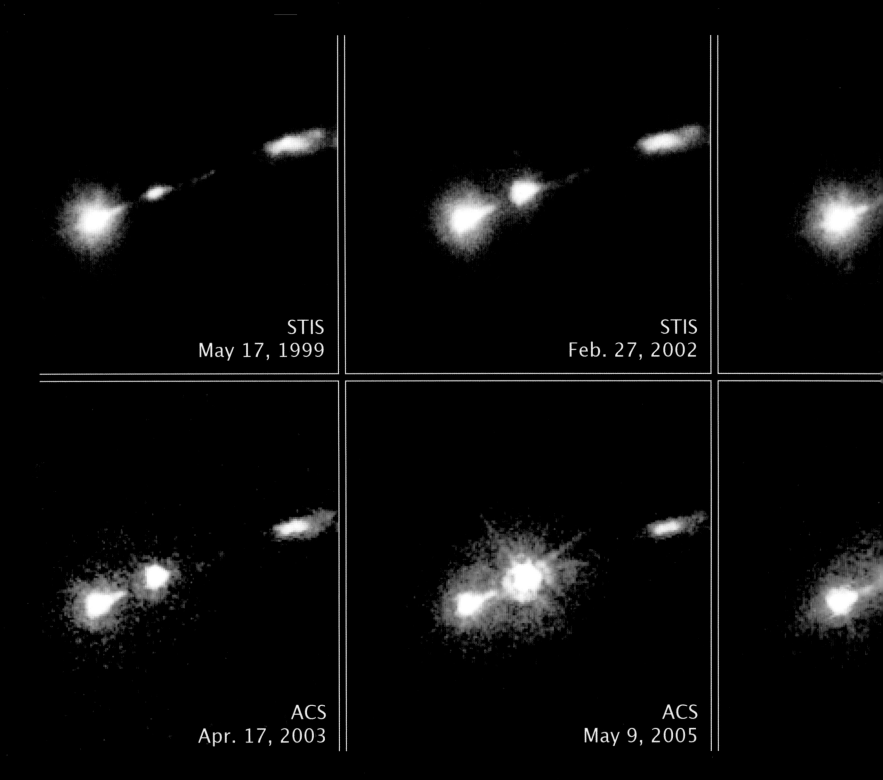

Above In a series of six pictures over seven years, the HST recorded a jet of gas from the black hole in the centre of the galaxy M87 as it shot into a knot of hot gas, causing it to flare up.

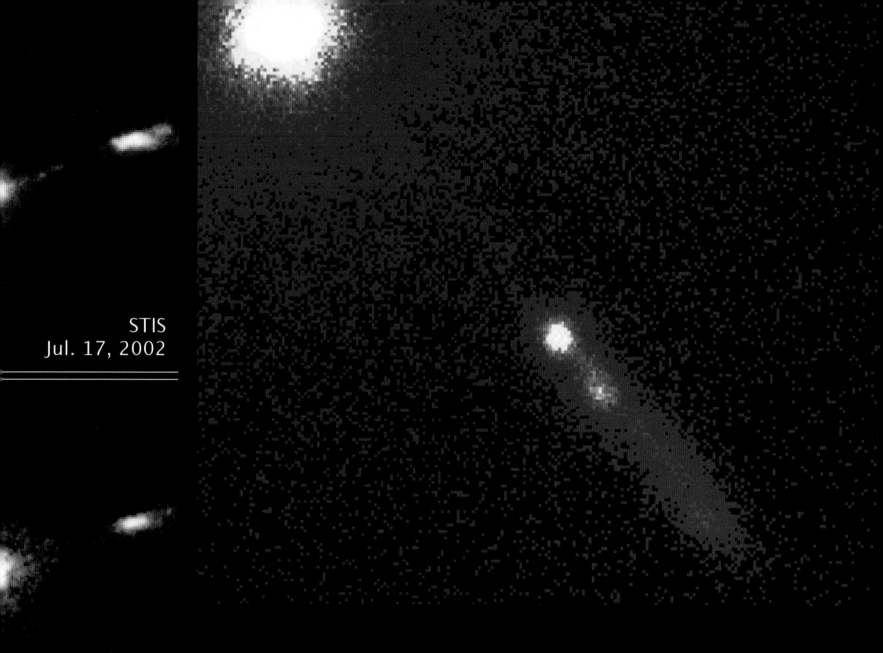

STIS
Jul. 17, 2002

ACS
Nov. 28, 2006

Above right *X-rays viewed by the telescope on the Chandra satellite show a jet that shoots away from the black hole in the distant galaxy and quasar, 3C273.*

3C273 – A GALAXY FAR, FAR AWAY

A strong radio source, known from its catalogue reference as 3C273, lies in the zodiac, which is the path of the Moon as well as the Sun. In 1962, the Anglo-Australian radio astronomer Cyril Hazard used the newly-built Parkes radio telescope in Australia to watch 3C273 as it disappeared behind the Moon on several occasions. He was able to pin down its position by plotting the edge of the Moon at the moment of its disappearances.

Right on the position of 3C273 was what looked like a star with a small wisp attached. CalTech astronomer Maarten Schmidt (1929–) investigated the "star", and discovered that it had a bizarre spectrum, unlike anything he had seen before. The answer hit him out of the blue. Some of the emissions were actually the very common, very familiar spectrum of hydrogen. The reason why Schmidt had not immediately recognized them was that the wavelengths were increased (redshifted) by a huge factor – 3C273 was taking part in the expansion of the universe and was a galaxy far away, not a star in our own Galaxy.

The expanding universe

In the first decade of the twentieth century, Lowell Observatory astronomer Vesto Slipher (1875–1969) engaged in a programme to investigate the composition of "spiral nebulae" at a time when it was not clear that they were galaxies outside our own. As a side product of his programme, in 1912 he measured the speed of the Andromeda Nebula, M31, the greatest speed of a celestial body hitherto observed. He went on to measure the speeds of over a dozen spiral nebulae and found that nearly all were receding away from us.

The Mount Wilson astronomer Edwin Hubble (1889–1953) extended Slipher's list with measurements he made with a colleague, Milton Humason (1891–1972). In 1929, Hubble demonstrated that there was a proportionality between the galaxies' distances and their recessional speed or redshift (see page 63). This correlation is now known as Hubble's Law. The paper was immediately noticed, because just before its publication, a Belgian mathematician, the Abbé Georges Lemaître (1894–1966) derived a mathematical solution to the structure of the universe from Einstein's General Theory of Relativity, predicting that the universe was expanding. If you are inside a uniformly expanding structure, the speed at which more distant objects recede from you is proportional to their distance. Together, the observational work and the theoretical calculation became the modern theory of the expanding universe.

Right *In 1904 the 60-inch Hale Telescope's tube structure was hauled by a mule train up a mountain track cut into the side of the San Gabriel mountains on its way to what was then the Mount Wilson Solar Observatory in California.*

Opposite *Sir Martin Ryle (right) and the then President of the Royal Society Sir Alan Hodgkin stand in 1972 in front of one of the antennae of the 5 km radio telescope in Cambridge, England.*

As exemplified by the work by Maarten Schmidt, many sources of celestial radio waves were being shown to be galaxies at great distances. The expansion itself became manifest in 1951, when the team of Cambridge radio astronomers led by Martin Ryle (1918–1984) showed how the fainter radio galaxies were packed closer together than the brighter ones, a discovery for which (in part) he was awarded a Nobel Prize in 1974. The interpretation of his work is that, because fainter radio galaxies are further away than brighter ones, and radio waves from them have taken a long time to reach us, the universe was more compact in the past than it is now. The universe had a beginning, an explosion which was continuing. Galaxies were even now expanding one from another.

Lemaître visualized the origin of the expansion of the universe as the explosion of a hot, small, dense "atom", a phenomenon which became known as the Big Bang. The implication is that at the origin of the universe everything was very hot, a mixture of matter and radiation at extremely high temperatures. Both still fill the universe, the matter having cooled and become galaxies. The Hubble Space Telescope was named after Hubble because the main scientific case for it was to make a definitive study of Hubble's Law, as a Key Project. Observations using it showed

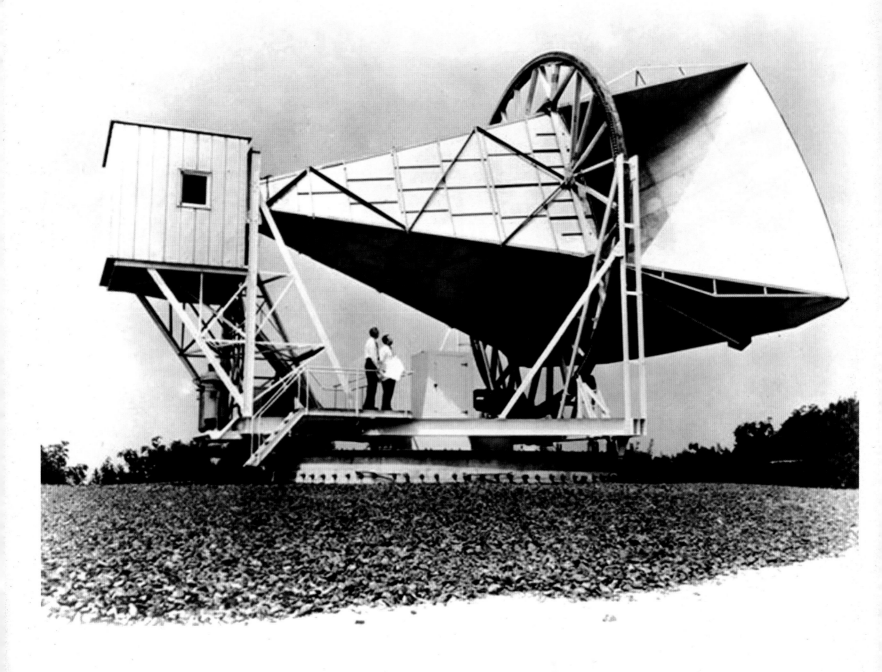

that the age of the universe, the time at which the explosion started, was 13.8 billion years ago. As the universe has expanded over the 14 billion years since, the wavelength of the radiation has lengthened, and the radiation itself has also become cooler. The United States physicists George Gamow (1904–68), Ralph Alpher (1921–2007) and Robert Herman (1914–97) calculated in 1948 that the radiation would now be at a temperature, perhaps, somewhere between 5 K above absolute zero and 28 K. Its wavelength would be typical of "microwaves" – radiation that could be described as very short wavelength radio waves or rather long wavelength infrared radiation.

The radiation has become known as the cosmic microwave background. It was discovered in 1965, as an incidental to a study by Arno Penzias (1933–) and Robert Wilson (1936–) of Bell Telephone Laboratories in New Jersey. They built a radio antenna to the most advanced specification that they could. It always showed a source of radio noise that they could never get rid of by making instrumental improvements. In the end, they concluded that the noise was due to a natural background. Its temperature was 2.75 K, and it was soon identified as cosmic microwave radiation. Penzias and Wilson were awarded the 1978 Nobel Prize in Physics for this discovery.

Above *Arno Penzias and Robert Wilson ride on the rotating turntable of the vast horn-shaped telescope in Holmdel, NJ, with which they discovered the Cosmic Microwave background.*

MILTON HUMASON

Milton Humason had had a remarkable career at the Mount Wilson Observatory, having left school aged 14. He started work there as a mule skinner during the Observatory's construction, hauling building materials up the mountain. He briefly left to work on a ranch, but rejoined the Observatory as a janitor in 1918. His qualities were recognized by the Observatory Director, George Ellery Hale (1868–1938), and he was appointed to the scientific staff, first as an operator for the 100-inch telescope and then as a research assistant to Hubble. He was known for his skill in operating that telescope and, while taking photographs of variable stars for Hubble, would hang on to particular parts of the cell that contained the mirror, because he had found that doing so compensated for some irregular mechanical flexures that distorted the mirror's images.

Left *Milton Humason, by turns over most of the lifetime of the Mt Wilson Observatory a mule skinner, a janitor and an astronomer.*

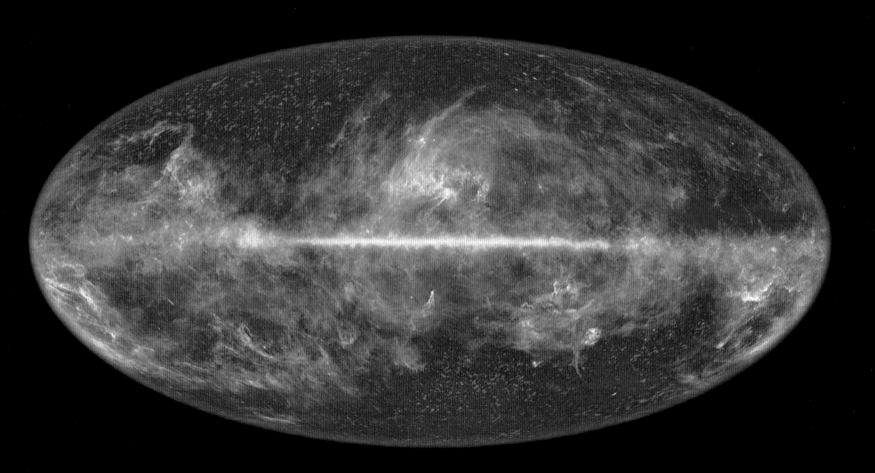

Above *The first all-sky picture of the Cosmic Microwave Background and the Galaxy made in 2010 by the Planck satellite. Red patches are brighter than blue, and the strip of sky along the Milky Way predominates. The sky above and below the Milky Way is the Cosmic Microwave Background, and its faint mottles represent the structures in the material of the Big Bang that grew into everything astronomers now think is interesting.*

ANISOTROPIES

The cosmic microwave radiation comes from everywhere and is very homogeneous ("isotropic"), but there is a limit to its isotropy that is set by the fact that although the Big Bang homogenized everything, it was not completely uniform – in fact, galaxies, stars and planets (and you and me) grew out of these tiny irregularities. The NASA satellite COBE (COsmic Background Explorer) measured the slight irregularities in the cosmic microwave background in 1989 at a level of one part in 100,000, an achievement recognized by the award of the Nobel Prize of 2006 to the project leaders John Mather and George Smoot. The follow-up NASA satellite WMAP (Wilkinson Microwave Anisotropy Probe) made even more accurate measurements in 2003, and an ESA satellite, Planck, in orbit from 2009, is expected to provide a definitive map of the irregularities in due course. The reason for all this effort is that the properties of the irregularities provide amazingly accurate information about the start of the universe, its composition and its future. The best fit interpretation is that cosmic microwave background originated only 380,000 years after the Big Bang, 13.8 billion years ago, and that the universe will expand for ever.

Dark matter and dark energy

Above *Abell 1689 is one of the most massive and populous clusters known, containing thousands of galaxies. Further thousands of galaxies lie behind this cluster, smaller because they are more distant, and distorted by its gravitational lens effect. They make elongated arcs that are centred on the most massive and brightest galaxy in the middle of the cluster, and provide the means by which the cluster's content of dark matter can be measured and mapped.*

Being humans, astronomers pay attention to things that they can sense, and for about 100 years they have thought of the universe as being populated by galaxies of stars that shine with light and gas that radiates radio and other waves, spread throughout a large volume of totally inert space that does not do anything except provide somewhere for all these things to exist. The stars are made, as we are (because we are a spin-off from stars), of "matter" - atoms, themselves made of protons, neutrons, electrons and other similar so-called elementary particles. "Matter" is, or so it has appeared for the last century, the main constituent of the universe.

The first indications that we are, in fact, actually ignorant about most of the content of the universe came in 1933 when CalTech astronomer Fritz Zwicky (1898–1974) noticed that galaxies that congregated in clusters of galaxies are moving too quickly – too quickly to be pulled by the gravity of all the stars in the other galaxies in the cluster. Zwicky speculated that there was some form of "missing matter" or "dark matter", something that did not radiate light as stars do, but which had mass and was able to augment the force of gravity of all the stars in a galaxy and speed up other galaxies as they moved past.

It was a radical idea and not initially widely accepted, but it gained credibility in the 1970s when Vera Rubin (1928–) found a similar phenomenon within galaxies. There was something that she could not identify that added to the mass of stars in a given galaxy, pulling the stars along more quickly than expected.

Zwicky had earlier indicated another line of investigation to determine how much mass there is in a galaxy, with the concept of a "gravitational lens". According to Albert Einstein's General Theory of Relativity, the mass of a galaxy distorts space around it and makes space into a distorting lens that can magnify the image of anything that lies beyond. If there is another galaxy in the background, its images in the lens can be displaced, distorted and made brighter. The nature of the image depends on the mass of the foreground lensing galaxy. This effect was first discovered as a real phenomenon by Dennis Walsh (1933–2005), Ray Weymann and student Bob Carswell in 1979. The mass of the lensing galaxy is always much too big to be accounted for only by its stars. There is typically five times as much mass in the extra "dark matter" than in the stars.

Since there is so much dark matter, it has a considerable effect on the expansion of the universe. The force of gravity acts between all the galaxies as they expand outwards, and tends to slow the expansion down. In an attempt to measure this, the Hubble Space Telescope was used by two large teams (the High-z Supernova Search Team and the Supernova Cosmology Project) in 1998–99 to study the expansion rate of distant galaxies compared with the expansion rate in our vicinity. If the expansion was slowing down, the expansion rate of distant galaxies (seen as they were in the distant past) should be larger than the expansion rate of ones close by.

It was a considerable shock to astronomers to find that the reverse was the case. The universe is expanding more quickly now than it used to. Something is accelerating the explosion of the Big Bang. For want of a better name, the "something" is called "dark energy", an energy released in considerable quantities by space itself into the expansion of the universe.

What we know about dark matter and dark energy (which is not much) has been factored into the study of the Cosmic Microwave Background (CMB). The small irregularities in the

CMB represent lumps in the material of the Big Bang which pull in material from their neighbourhood and grow into galaxies. Dark matter adds to the gravity of the lumps and pulls in larger amounts of material more quickly than if there was no dark matter. On the other hand, dark energy acts as a kind of repelling force and slows down the condensation of the lumps. A calculation with the most powerful modern supercomputers of the way that this works was published in 2005, and is known as the Millennium Simulation. It progressively followed the tracks of 10 billion particles, representing galaxy-sized lumps. When the calculation was matched to observations of the CMB made by the satellites COBE and WMAP, astronomers concluded that only 4 per cent of the universe is ordinary matter: 21 per cent is dark matter; and a whopping 75 per cent is dark energy, both of them in unknown form. After 100 years studying modern cosmology, astronomers have found out that they are ignorant about 96 per cent of the content of the universe.

Above *The central part of the Virgo cluster of galaxies.*

Below *The Large Hadron Collider at CERN in Geneva may be able to detect and identify the particles that might make up dark matter, orbiting galaxies in such abundance that, as Vera Rubin discovered, they control the very orbits of stars. Note the technician making an adjustment at the centre of this gigantic particle detector.*

VERA RUBIN

Attracted to astronomy as a child on observing what to her was a fascinating fact, that the stars rotated around the celestial pole, Vera Rubin graduated from Vassar. Having overcome difficulties in her early career, because at the time women were not welcomed into astronomy at the large departments at many prestigious universities, Rubin joined the Carnegie Institution in Washington D.C., which has a modest astronomy programme. Pairing up with instrumentalist Kent Ford (1931–), who had developed an extremely sensitive instrument with which it was possible to study the rotation of stars around galaxies, she found in the 1970s that, just as galaxies were moving too fast in clusters of galaxies, stars also were moving too fast within the galaxies themselves.

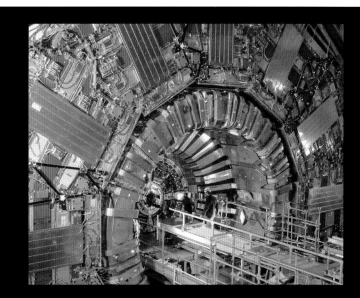

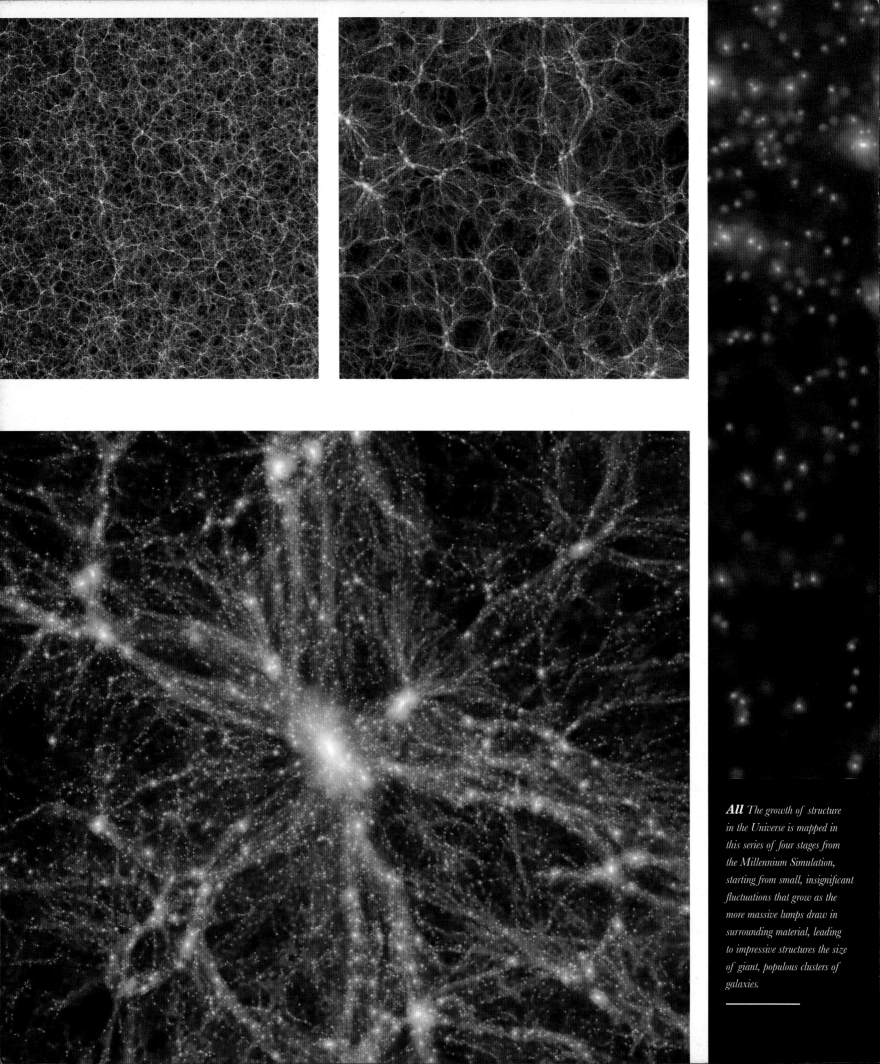

All *The growth of structure in the Universe is mapped in this series of four stages from the Millennium Simulation, starting from small, insignificant fluctuations that grow as the more massive lumps draw in surrounding material, leading to impressive structures the size of giant, populous clusters of galaxies.*

WHAT IS DARK MATTER?

The first idea about the dark matter was that it might be inert gas or dead stars – white dwarfs, neutron stars, black holes. But this proved not to be the case, and astronomers have turned to the Big Bang for an alternative explanation. The Big Bang made all the matter in the universe and it may have made dark matter particles too. These particles could be of a type unknown to science. Presumably they do not interact much with ordinary matter, or else we would have noticed them already, so they go by the sly acronym WIMPs (for Weakly Interacting Massive Particles). They have to be very massive, or else particle physics accelerators would have made some already and that would have been noticed too. It is possible that, in the future, more powerful accelerators than have yet been constructed will have the energy to make and detect some of them. This is one of the aims of the so-called Large Hadron Collider accelerator (LHC) in Geneva, which began operation in 2008.

30

Life in the Universe

Is there life elsewhere in the Universe or are we alone? The first discussions were philosophical – if life is a natural phenomenon, it could exist anywhere that conditions allow. "There is an infinite number of worlds, some like this world, others unlike it," wrote the Greek philosopher Epicurus (341–270 BC), "… in some sorts of world there could be the seeds out of which animals and plants arise". Giordano Bruno (1548–1600), expressed the same opinion in his book *De l'infinito universo e mondi* (1584): "The countless worlds in the universe are no worse and no less inhabited than our Earth."

Philosophy shaded into a new science of astrobiology with the proof that the elements critical to life, principally carbon, are made in stars, available everywhere, and that planets are indeed common. Of course, conditions matter – some planets are habitable, others not. In 1953, astronomer Harlow Shapley (1885–1972) identified the "liquid water belt" in the Solar System, the zone in which liquid water could persist on a planet, and support life. With the wrinkle that this is not just a matter of distance from the warmth of the Sun (as shown, for example, by the ocean of water on Jupiter's freezing satellite Europa), the concept of the "Goldilocks zone" became the guide for a search by NASA for life on other planets; within this zone the temperature is not too hot that water turns to steam, not too cold that it freezes, but just right.

If there is life on another planet, could we communicate with it? A century ago, early radio pioneers like Guglielmo Marconi (1874–1937) thought that they had picked up transmissions from outer space, possibly from Martians. From 1960, systematized but still fruitless searches, listening for intelligent signals, have been implemented under the name of SETI, the Search for Extraterrestrial Intelligence, first directed by radio astronomer Frank Drake (1930–). SETI continues and anyone can enrol in "SETI@home", a project in which your home computer is used to analyze radio signals for artificial messages.

The SETI project estimates that there are numerous civilizations in our Galaxy that would like to talk to us. This may be optimistic. Under the name "Rare Earth", some modern theories have identified lucky flukes in the history of our Earth that have given our planet properties uniquely favourable for intelligent life, such as stability for long enough for it to develop.

If the "Rare Earth" theories are right, life may be ubiquitous in the universe but intelligent life may be rare. Two thousand years of astronomical history have not yet been enough to decide the question.

Above *The massed stars of the Milky Way (here the star clouds of the centre of our Galaxy) are so numerous, all of them suns of one kind and another, many of them with retinues of planets, that it is almost compelling to believe that some of them must be the home to creatures like us in some way.*

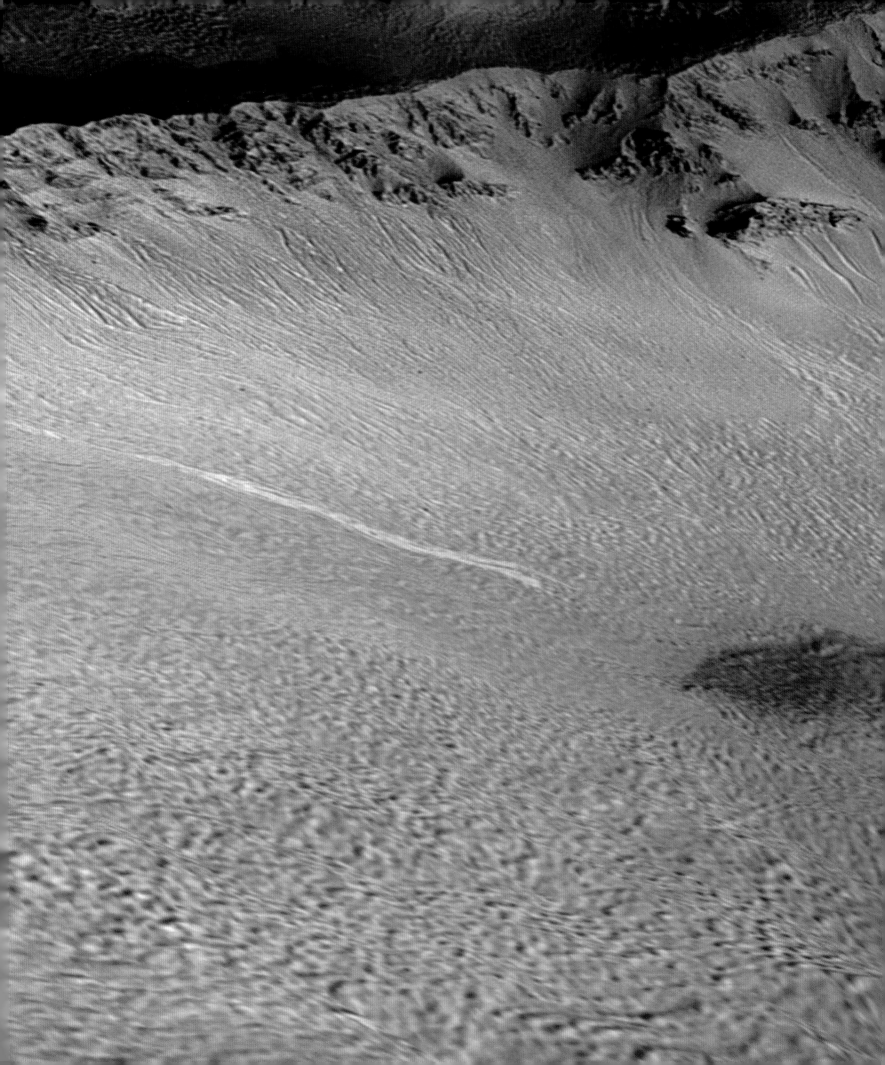

THE DRAKE EQUATION

In 1966 Soviet radio astronomer Iosif Shklovsky (1916–1985) and United States planetary scientist Carl Sagan (1934–1996) calculated the number of habitable planets. Their calculations were drawn together by Frank Drake into the "Drake Equation". He expressed the number, N, of communicating civilizations in the Galaxy through a number of factors:

$$N = N_g\, f_p\, n_e\, f_l\, f_i\, f_c\, f_L$$

Here N_g is the number of stars in our Galaxy, a fraction f_p of which have planets, n_e earths per planetary system; f_l is the fraction of earths on which life evolves, f_i the fraction which produce intelligent life, f_c the fraction of civilizations with radio telecommunication ability, and f_L the fraction of a planet's lifetime during which it can support life.

Sagan "guesstimated" the number of radio-transmitting civilizations in the Galaxy at perhaps one million.

Opposite *In 2004 the Mars Global Surveyor spacecraft found a new feature in a crater in the Centauri Montes region of Mars, which had appeared since August 1999. A white branching stain had appeared on the wall of the crater, apparently sediment carried by a geothermal spring of water that had temporarily gushed from the wall, draining downhill.*

Above *SETI@home is a citizens' science project that uses the power of the large number of home computers in the world's population to search for interstellar radio messages. It could be your computer that brings up evidence of a communication from extraterrestrial aliens for your eyes to be the first to see, on your computer-screen!*

Index

Index

Illustrations are indexed by the page on which the caption appears.
Ch: chapter

3C273 140–3
61 Cygni 86

A
Abt, H., 8
accretion 125
Adams, J.C., 59
Adams, W. S., 103
Airy, G., 59
Albategnius (crater) 109
al-Din ibn Maruf, Taqi, 27
Aldrin, B., 98
Algol 71
al-Ma'mun 27
Alpher, R., 146
al-Rashid, H., 27
al-Sufi, Abd al-Rahman, 27, 132, 139
Anders, W., 108
Andromeda Nebula (Galaxy) 130, 132, 134, 144
Antikythera mechanism 21
Apollo 102
Apollo 8 108
Apollo 11 98
Apollo 15 72
Apollo programme 98
Aquinas, T., 24
Arago, F., 59
Aratus 14, 17
Aristotle 18, 20, 24, 46
Arizona Meteor Crater 108
armillary sphere 12
asterism 14
asteroids 58, 97
astrolabe 13
astrology 17
  and medicine 13
astronomical instruments, pretelescopic, Ch 2

Atkinson, R. d'E., 74,
Averroes 23,

B
Baade, W., 86, 140
barred spirals 137, 138
Barringer, D., 108
Barstow, M., 81
Becklin, E., 94
Beijing Observatory 12
Bell, J., 86
BepiColombo (spacecraft) 104
BeppoSAX (spacecraft) 88
Bessel, F., 66
Beta Pictoris 94
Betelgeuse 78
Bethe, H., 74
Big Bang 144
Big Dipper 8, 17, 18, 24
black hole 81, 86–7, 140–3, 151
Blake, W., 48
Bloxam, R.R., 17
BN Object 94–5
Bode, J., 56
Bode's law 56
Borman, F., 108
Bruno, G., 124, 152
Buffon, Comte, 72
Bunsen, R., 66
Burbidge, G., 91
Burbidge, M., 91
Burke, B., 116
Butterfly Nebula 78

C
calendar 26
Callisto 119
Caloris basin 104
Cameron, A., 91
Carswell, R., 148
Cassini, G.D., 54, 110, 116
Cassini (spacecraft) 116, 121
Cassiopeia A 74
Cat's Eye Nebula 78
Cellarius, A., 18, 30

Cepheid variable stars 130
Ceres 58
Chaco Canyon, NM, 7–9, 9
Chandra X-ray telescope 46, 80–2, 91, 142
Chandrasekhar, S., 78, 81
Chandrasekhar Limit 81
Chinese astronomy 12, 17
chondrite meteorite 97
Christiansen, W., 131
chromatic aberration 54
Clark, A., 54
Colombo, G., 102
Columbus, C. 22
Comet McNaught 62
Comet Shoemaker-Levy 9 116, 119
Comet Wild 2 64
comets 64, 97
constellations Ch 3
Coordinated Universal Time 53
Copernicus, N., Ch 6, 45, 108, 124
CoRoT (spacecraft) 125
Corsali, A., 139
Cortés de Albacar, M., 24
COsmic Background Explorer (COBE) 147
Cosmic Microwave Background 146–7
CP1919 (pulsar) 86
Crab Nebula 88
Crabtree, W., 44
craters 97
  (on Earth) 108
  (on Europa) 119
  (on Moon) 108
  (on Mars) 104, 110, 114
  (on Mercury) 102, 41
Creti, D., 40, 110–11
cross-staff 12
cryovolcanoes 123
Curtis, H., 130
Cusco, Peru, 9
Cygnus A 108-9, 86
Cygnus X-1 82
Cysat, J.-B., 93

D
Daly, R., 108
Dark Ages Ch 5
dark energy Ch 29
dark matter Ch 29, 151
de Peiresc, N., 93
Deimos 110
Democritus 22
Dixon, J., 107
Dollond, J., 54
Drake, F., 155
Drake equation 155
Draper, H., 137
Dresden Codex 9
Dreyer, J.L., 132
Dunhuang star chart 17, 18
dwarf galaxies 138

E
Eagle lunar lander 98
Earth Ch 22, 102
  age, 6
  liquid core, 108
  magnetic field, 98
  orbit, 66
  shape, 22
  structure, 91
  tectonic activity, 108
Eddington, A.S., 69, 74, 81
Edgeworth, K., 61
Edgeworth-Kuiper Belt 62
Einstein, A., 76, 81, 87, 104, 144, 148
Einstein (spacecraft) 62
El Castillo 9
electron degenerate pressure 78
elements Ch 18
  (earth, air, fire and water) 20
elliptical galaxies 137
Empyreum 24
Enceladus 121, 123
Eosphoros 102
Epicurus 152
Eratosthenes of Cyrene 22
Eris 62
Eskimo Nebula 78, 81

Eudoxus of Cnidus 14
Europa 119, 152
event horizon 87
Ewen, H., 131
expanding universe Ch 28
exploration of space Ch 20
  Grand Tours, 101
  robotic, 100
Explorer (spacecraft) 98
extrasolar planet Ch 24, 94

F
Fabian, A., 140
Fabricius, D., 69
Farnese Atlas 14, 15
Firmament 24
Flammarion, C., 20, 8
Flamsteed, J., 49
Fomalhaut 124
Ford, H., 140
Ford, K., 148
Fowler, R., 78
Fowler, W., 90–1
Frail, D., 124
Franklin, K., 116
Fraunhofer, J., 66

G
G292.0+1.8 (supernova remnant) 91
Gagarin, Y., 108
galaxies Ch 25–6
  dwarf, 138
  elliptical, 137
  irregular, 137
  mergers, 138
  spiral, 138
Galaxy Ch 25
Galileo Galilei Ch 8, 54, 108, 121,
  and the Church 39
Galileo (spacecraft) 98, 111
Galle, J., 59, 68
gamma ray astronomy 86, 90
Gamma Ray Observatory (spacecraft) 88

gamma-ray bursts 86–8,
Gamow, G., 146
Ganymede 119
gas giant planets Ch 23
General Theory of Relativity 81, 87, 104, 144, 148
Genesis Rock 72
Giacconi, R., 82
Giza, Egypt, 8
Gomes, R., 97
Good, M., 85
Goodricke, J., 71
Gran Telescopio de Canarias 55
gravitational lens 125, 148
gravitational redshift 81
gravity Ch 10,
gravity assists 101
GRB 080319B 86–7
GRB 970228 88
GRB 970508 88
Great Bear 14–17, 18
Great Red Spot 40, 116
greenhouse effect 107
Greenstein, J.L., 81
Greenwich Mean Time 50–3
Greenwich meridian 53
Gregorian calendar 26

H
Hall, A., 110
Hall, C.M., 54
Halley, E., 48–9, 65
Halley's comet 48, 49, 50
Harriott, T., 34
Hawkins, G., 6
Hayashi, C., 94
Hazard, C., 143
heliometer 66
Helmholtz, H., 72
Herman, R., 146
Hermes 102
Herschel, C., 56
Herschel, J., 132
Herschel, W., 56–7, 68, 78, 94, 126, 132
Hertzsprung, E., 68
Hertzsprung-Russell Diagram 68, 78
Hesperos 102
Hewish, A., 86
Hey, J., 82
Hindman, J., 131
Hipparchus 13, 21
Holbein, H., 24
Hooke, R., 116
horoscope 17
Horrocks, J., 43
Houtermans, F., 74
Hoyle, F., 6, 91

Hubble, E., 137–8, 106, 110
Hubble Deep Field 71, 138
Hubble Space Telescope 82, 85, 88, 94–5, 89, 130, 145
Hubble Ultra Deep Field 71, 138
Hubble's Law 144–5
Huggins, W., 66
Humason, M., 144, 147
Huygens, C., 54, 110, 121
Huygens Lander 121
hydrogen bomb 76
hypernova 87

I
Inca astronomy 9
infrared astronomy 82–5, 94
InfraRed Astronomy Satellite (IRAS) 94
inner planets Ch 21
International Atomic Time 53
Io 116, 119
irregular galaxies 137
Irwin, J., 72
Islamic astronomy 26

J
Jacob's staff 12
Jai Singh II 13
James Webb Space Telescope 85
Jansky, K., 84
Jantar Mantar, India, 13
Jewitt, D., 62
Johannes de Sacrobosco 24
John of Holywood 24
Joint European Torus 74
Julian calendar 26
Julius Caesar 26
Juno 58
Jupiter Ch 23, 34, 40
  atmosphere, 116
  formation, 97
  magnetic field, 116
  radio noise, 116
  satellites 34, 40, 119
  structure, 116

K
Kant, I., 94, 130, 137
Kapteyn K. 126
Karnak, Thebes, 8
Keck Telescope 55, 88
Keeler, J., 121
Kelvin, Lord, 72
Kepler, J., Ch 9, 29, 41
Kepler (space mission) 125
Kepler's laws of planetary motion 44, 48

Kepler's supernova 46
Kirchhoff, G., 66
Kuiper, G., 62
Kuiper Belt 97

L
Lady's Comet 56
Laplace, P.S., 87, 94, 121, 137
Large Hadron Collider 148, 151
Large Magellanic Cloud 90, 127, 139
Le Verrier, U., 59
Leclerc, G.L., 72, 145
Lemaître, G., 144
Leucippus 22
Leverrier, U., 104
Leviathan of Parsonstown 86, 135
Levison, H., 97
Liceti, F., 121
Life in the Universe Ch 30
Lipperhey, H., 34
Local Group of galaxies 132
Lockyer, N., 9
Lomonosov, M., 104
Lovell, J., 108
Lowell, P., 61, 110
Lowell Observatory 61, 110
Luna (spacecraft) 98
Lutetia 62
Luu, J., 62

M
M31 (galaxy) 130–1
M51 (galaxy) 132
M87 (galaxy) 140
Maat Mons 105
Maddox Brown, F., 43
Magellan, F., 139
Magellan (spacecraft) 134–9
Magellanic Clouds 132, 139
Main Sequence 69
Maraldi, J.P., 110
Marconi, G., 152
Mariner 2 98, 104
Mariner 4 98, 112
Mariner 9 112
Mariner 10 104
Mars Ch 24, 110
  canali, 110
  exploration, 98
  moons, 110
  orbit, 29, 43, 42
  polar caps, 110, 112
  Syrtis Major, 110
  water, 112–15, 152
Mars Express (spacecraft) 110, 112

Mars Express Orbiter (spacecraft) 112
Mars Global Surveyor (spacecraft) 112, 155
Mars Odyssey (spacecraft) 112
Mars Pathfinder (spacecraft) 98–9, 112
Mars Reconnaissance Orbiter (spacecraft) 112
marsh gas 114
Martianus Capella 18
Mason, C., 107
Massimino, M., 85
Mather, J., 147
Maxwell, J.C., 121
Mayan astronomy 9
Mayor, M., 125
MCG-6-30-15 (galaxy) 140
Mercury 107
Messenger (spacecraft) 104
MESSENGER spacecraft 104, 132
meteorite 97
Michell, J., 87
Milky Way 33, 126–31
Milky Way, radio noise, 82
Millennium Simulation 149
Miller, W., 8
Mimas 121, 123
Mira 69
Miranda 125
Montanari G. 71
monuments Ch 1
Moon Chs 23, 28
  eclipse, 22
  exploration, 98
  origin, 97
  X-rays from, 82, 37
moonquakes 98
Morabito, L., 119
Morbidelli, A., 97
Muhammed ibn Rushd 27
Muller, A., 131

N
Nazca Lines 9
nebulae 93, 131
Nebular Hypothesis 94, 137
Neptune 59, 61, 123
  orbit, 61
Neugebauer, G., 94
neutron star 86, 91, 151
New Horizons mission 101
Newcomb, S., 72
Newgrange, Ireland, 8
Newton, I., Ch 10, 55, 124
NGC 1132 (galaxy) 137
NGC 1514 (planetary nebula) 78

NGC 2346 (planetary nebula) 78
NGC 4526 (spiral galaxy) 87
NGC 5195 (galaxy) 132
Nice Model of solar system 97
nocturnal 24

O
O'Dell, R., 94
Olbers, Heinrich 45
Olympus Mons 112
Omicron Cet 69
Oort, J., 64, 131
Oort Cloud 64, 97
Opportunity (spacecraft) 112
Orion 9, 41, 66, 93
Orion Nebula 93–7
Owl Nebula 78, 81

P
Pallas 58
parallax 46, 66
Paris Observatory 53
Parsons, W., 132
Parsons, W., 135
Payne-Gaposchkin C. 66
Penzias, A., 146
Peurbach, G. von, 28
Phobos 110
Phoenix 112, 68
Piazzi, G., 58
Pickering, E., 66
Pioneer 10 (spacecraft) 116
Pioneer 11 (spacecraft) 121
Pitluga, P., 9
Planck (spacecraft) 147
Planet X 61
planetary nebula 78, 91, 94
planetary system, Goldilocks zone, 152
  liquid water belt, 152
planetesimals 97
planets, birth, Ch 19
  exploration, 24
Plough 8, 17, 24
Pluto 61, 97, 101
Pointers 8, 24
Polaris 24
Pope Gregory XIII 26
Potsdam Astrophysical Observatory 69
Pound, R., 81
Prime Meridian 53
prime mover 20, 24
proplyd 94, 97
proto-planetary systems 94
PSR 1257+12 125
Ptolemaeus, C., 17
Ptolemy 23

pulsar 86
pulsar planets 125
Purcell, E., 131
pyramids 8

Q
qiblah 27
quadrant 13, 30
quasars Ch 27, 140
Queloz, D., 125
quintessence 20

R
radio astronomy Ch 16, 131
  140
radioactive thermoelectric
  generators (RTGs) 101
Rare Earth 152
Reber, G., 82
Rebka, G., 81
red giant 78
red supergiant 78
reflecting telescope 55
refracting telescope 54
Reiche, M., 9
Reifenstein E. 86
Rheticus 28
Ring Nebula 78
Robert-Fleury, J.-N., 39
Roberts, I., 137
Roche, E., 121
ROSAT (spacecraft) 82
Rosetta (spacecraft) 62
Royal Observatory
  Greenwich 50–1
Rubin, V., 148
Russell, H.N., 68, 78, 145
Rutherford, E., 72
Ryle, M., 144

S
S Andromedae 130
Sagan, C., 155
Sagittarius dwarf galaxy 138
Samarkand 27
Saturn 37, 121, 145
  rings, 121, 145
Schiaparelli, G., 110
Schmidt, M., 143, 86
Schwarzschild, K., 87
Scorpius X-1 82
Scott, D., 72
Search for Extraterrestrial
  Intelligence (SETI) 152
Secchi, A., 66, 110
SETI@home 152, 130
sextant 13
Shapley, H., 126, 152
Shklovsky, I., 155
Shoemaker, E., 108

Signs of the Zodiac 26
Sirius 81
Slipher 144
Small Magellanic Cloud
  127, 139
Smith, F.G., 140
Smoot, G., 147
Soddy, F., 72
Sojourner rover 100
Solar Max (spacecraft) 90
solar system 94
  Copernican, 28, 41
  heliocentric, 28
  Kepler's model, 42
  Tycho's model, 18, 30
  geocentric model, Ch 4
Somerville, M., 59
Southern Cross 14
spectral lines 66
spiral galaxies 137–8
spiral nebulae 137–8
Spirit (spacecraft) 112
Spitzer (spacecraft) 93
Sputnik (spacecraft) 98
Staelin, D., 86
Star of Bethlehem 46
stars Ch 13
  birth, Ch 19
  colour, 66
  death, Ch 15
  distance, 66
  dwarf, 90
  dwarfs, 69
  eclipsing, 71
  giants, 69
  internal constitution, 69
  lives, Ch 14
  luminosity, 66
  mass, 68
  nuclear fusion, 72–6
  spectroscopy, 66
  supergiants, 69
  temperature, 66
Stjerneborg 30, 32–3
Stonehenge 6
Stukeley, W., 6
Subaru telescope 55
Sun, age, 72
  destiny, 74
  nuclear energy, 74, 82
sundial 24
sunspots 78
supergiants 90
supernova Ch 17, 86, 90–1,
  148
Supernova 1994D 87
Supernova of 1054 8, 46, 86
Supernova of 1572 30, 43
Supernova of 1604 46
Supernova of 1987 90

supernova remnant 46, 91
Swedenborg, E., 94, 137
telescopes Ch 11
  Galileo's, 34–7
  Herschel's, 56

T
terrestrial planets 97
Thales of Miletus 20
Theia 108
Thomson, W., 72
timekeeping 24
Titan 121, 123
Titius, J., 56
Tolosani, G. M., 29
Tombaugh, C., 61
Trans-Neptunian Objects 62
Trapezium stars 93
Trifid Nebula 97
Triton 123
Tsiganis, K., 97
Tuning Fork Diagram 138
Twin Peaks (on Mars) 100
Tycho Brahe Ch 7, 42, 46
Tycho's supernova 43

U
UHURU (spacecraft) 82
Ulugh Beg 27
Uraniborg 30, 32
Uranus 56–8, 123
  orbit, 58
Ursa Major 14–17
Uxmal, Mexico, 9

V
Van Allen 98
van de Hulst, H., 131
van Ghent, J., 18
van Leeuwenhoek, A., 15
Vega 94
Vela (spacecraft system) 86
Venera 13 102
Venera 14 101
Venera (spacecraft) 98
Venus 41, 100, 102–04
  atmosphere, 104–07
  calendar, 9
  exploration, 98
  orbit 41
  phases, 37
  transit, 43, 107
Venus Express mission 107
Verbiest, F., 12
Vermeer, J., 15
Very Large Array (VLA) 140
Very Large Telescope
  (VLT) 55
Vesta 58
Viking 1 and 2 102

Viking Landers 68, 101
Virgo cluster of galaxies 148
Vogel, H., 66
Voyager 1 101, 121
Voyager 2 101, 121, 123
Voyager (spacecraft) 121
Vulcan 107

W
Walsh, D., 148
Weakly Interacting Massive
  Particles (WIMPs) 151
Wegener, A., 108
Weird Terrain 104
Weizsäcker, C. von, 74
Wells, H.G., 110
Wesley, A., 119
Weymann, R., 148
Whirlpool Galaxy 132, 135
white dwarf 69, 78, 81, 86,
  151
Wilkinson Microwave
  Anisotropy Probe (WMAP)
  13
Williams, R., 138
Wilson, R., 146
Wolszcan, A., 124
Wren, C., 53
Wright, T., 126–30

X
XMM-Newton X-ray
  telescope 46, 84
X-ray astronomy Ch 16, 78,
  82, 86, 91

Y
Yerkes Telescope 55
Yuty (Mars crater) 114

Z
Zach, F.X. von, 58
Zeta Leporis 94
zij 27
zodiac 18–19
Zwicky, F., 86, 148

Picture credits

6-7 Istockphoto.com, 8 Alamy/George H.H. Huey, 8–9 Getty Images, 9br Corbis/Macduff Everton, 10–11 Corbis/Bob Krist, 12t Corbis/Bettmann, 12b Lonely Planet Images, 13t Getty Images/SSPL, 13b Corbis/Atlantide Phototravel, 14–15 Science Photo Library/Shelia Terry, 15 Getty Images/SSPL, 14 Corbis/The Gallery Collection, 15t Getty Images, 16 Private Collection, 17 Alamy/The Art Gallery Collection, 20 RMN/Hervé Lewandowski, 21t Private Collection, 21b Private Collection, 22tl Corbis/Antar Dayal/Illustration Works, 22–23 Science Photo Library/Dr. Fred Espenak, 23br Akg-Images/North Wind Picture Archives, 24bl The Bridgeman Art Library, 24–25 The Bridgeman Art Library/National Gallery, London, UK, 26 Getty Images/De Agostini, 27t The Bridgeman Art Library/Bibliotheque Nationale, Paris, France, 27b Science Photo Library/NYPL/Science Source, 28t The Bridgeman Art Library/Nicolaus Copernicus Museum, Frombork, Poland, 28b Science Photo Library/Royal Astronomical Society, 29l Science Photo Library/Shelia Terry, 29r Private Collection, 30–31 Science Photo Library/Royal Astronomical Society, 31 Science Photo Library/Royal Observatory, Edinburgh, 32t The Bridgeman Art Library/Private Collection, 32–33 Getty Images/SSPL, 34 The Bridgeman Art Library/Galleria Palatina, Palazzo Pitti, Florence, Italy, 35l The Bridgeman Art Library/Gianni Tortoli, 35r Special Collections Library, University of Michigan, 36 History of Science Collections, University of Oklahoma Libraries, 37tr Private Collection, 37b Private Collection, 38–39 The Bridgeman Art Library/Louvre, Paris, France, 40 The Bridgeman Art Library/Vatican Museums and Galleries, Vatican City, Italy, 41 Photo Scala, Florence, 42–43 Science Photo Library/Detlev van Ravenswaay, 43tr & r Topfoto.co.uk/World History Archive, 44 The Bridgeman Art Library/Manchester Art Gallery, UK, 44t Private Collection, 45 Science hoto Library/Marty Snyderman, Visuals Unlimited, 46–47 Science Photo Library/Caltech Archives, 46t NASA/CXC/Rutgers/J.Warren & J.Hughes et al, 46c NASA/ESA/JHU/R.Sankrit & W.Blair, 48–49 © Tate, London 2009, William Blake 1757–1827, colour print finished in ink and watercolour on paper, support: 460 x 600 mm, on paper, unique, 50t, 50b & 51t Getty Images, 51b Science and Society Picture Library/Science Museum Pictorial, 52 The Bridgeman Art Library/Private Collection, 53 Photolibrary.com/Japan Travel Bureau, 54 Private Collection, 55t Corbis/Jim Sugar, 55b Getty Images/SSPL, 56–57 Getty Images/SSPL, 56b Science Photo Library/Caltech Archives, 57 & 58–59 Science Photo Library/Royal Astronomical Society, 59t Getty Images/SSPL, 59b NASA, 60 NASA, 61 NASA, ESA, and M. Buie (Southwest Research Institute), 62t Getty Images, 62–63 Science Photo Library/Robert McNaught, 64r NASA/JPL/Stardust team, 65t The Bridgeman Art Library/ Musée de la Tapisserie, Bayeux, France, 65b Science Photo Library, 66t Science Photo Library, 66b Science Photo Library/Dr Jeremy Burgess, 67 © Akira Fujii/DMI, 68 Getty Images/SSPL, 68bl Science Photo Library/Library of Congress, 68–69 Alamy/LOOK Die Bildagentur der Fotografen GmbH, 70 Science Photo Library/Eckhard Slawik, 71t Science Photo Library/Mark Garlick, 71b NASA/Robert Williams and the Hubble Deep Field Team (STScI), 72t NASA, 72b Corbis/Bettmann, 73 NASA, 74–75 Science Photo Library/James King-Holmes, 74 istockphoto.com, 76–77 NASA and The Hubble Heritage Team (STScI/AURA), 76 United States Department of Energy, 78t Xavier Haubois/Observatoire de Paris et al, 78b, 79 & 80 NASA/AURA/STScl, 81l NASA, 81r NASA/SAO/CXC, 84t Image courtesy of NRAO/AUI, 84b NASA/ROSAT, 84–85 & 85 NASA, 86t Science Photo Library/Hencoup Enterprises Ltd., 86c Science Photo Library, 86b National Radio Astronomy Observatory, 87tl NASA/ESA/The Hubble Key Project Team, and The High-Z Supernova Search Team, 87 NASA/Swift/Stefan Immler, 88 Science Photo Library/Royal Astronomical Society, 88–89 NASA/Jet Propulsion Laboratory, 90t NASA/ESA and R. Kirshner (Harvard-Smithsonian Center for Astrophysics), 90b Science Photo Library/© Estate of Francis Bello, 91 NASA/CXC/Rutgers/J.Hughes et al, 92–93 NASA/ESA/M. Robberto (Space Telescope Science Institute/ESA) and the Hubble Space Telescope, 94t Science Photo Library/Deltev van Ravenswaay, 94b ESO/A.-M. Lagrange et al, 95 NASA/Rodger Thompson/Marcia Rieke/Glenn Schneider/Susan Stolovy (University of Arizona) Edwin Erickson, 96 NASA/Mark McCaughrean (Max-Planck-Institute for Astronomy) and C. Robert O'Dell (Rice University), 97t NASA/ESA and the Hubble Heritage Team (STScl/AURA), 97b Private Collection, 98 NASA, 100t NASA/JPL, 100b Russian Academy of Sciences, 101 NASA/JPL/University of Arizona, 102 Science Photo Library/Ria Novosti, 102–103 © Josselin Desmars, 104t NASA/JHU/APL, 104–105 & 106 NASA/JPL, 107 Science Photo Library/Royal Astronomical Society, 108 Science and Society Picture Library/Science Museum, 109 NASA/HQ/GRIN, 86 Science Photo Library/Detlev van Ravenswaay,

110t Photo Scala, Florence, 111b European Space Agency, 112, 112–113 & 114 NASA, 115 NASA/Hubble Site, 116 Science Photo Library, 116-117 & 118 NASA/JPL/Cornell University, 119t NASA, 119c NASA/JPL/University of Arizona, 119b NASA/Hubble Shoemaker-Levy, 120 Science Photo Library/US Geological Survey/NASA, 121t Science Photo Library/Royal Astronomical Society, 121b NASA/JPL/Space Science Institute, 122 Cassini Imaging Team/SSI/JPL/ESA, 123c NASA/JPL/SSI/LPI, 123bl Science Photo Library/NASA, 123br NASA/JPL, 124 NASA, ESA and P. Kalas (University of California, Berkeley, USA), 125t COROT exo-team, 125c & 125b EROS, 126–127 Science Photo Library, 126t Science Photo Library/Royal Astronomical Society, 128–129 ESO/H.H.Heyer, 127 History of Science Collections of the University of Oklahoma, 130–131 NASA/Swift/Stefan Immler (GSFC) and Erin Grand (UMCP), 131r Science Photo Library/Max-Planck-Institut Fur Radiostronomie, 132l Science Photo Library/Royal Institution of Great Britain, 132–133 NASA/ESA/S. Beckwith (STScI) and the Hubble Heritage Team (STScI), 133br Science Photo Library, 134 & 135l Science Photo Library/Royal Astronomical Society, 135 NASA, 136 NASA/ESA and the Hubble Heritage (STScI/AURA)/ESA/Hubble Collaboration. M. West (ESO, Chile), 137 NASA/ESA/S. Beckwith (STScI) and the HUDF Team, 138c Getty Images/Time & Life Pictures, 138b Science Photo Library/Royal Astronomical Society, 139 © Wei-Hao Wang (IfA, U. Hawaii), 140–141 NASA, 142 NASA/ESA/and J. Madrid (McMaster University), 143 NASA/CXC/SAO/H.Marshall et al, 144–145 Science Photo Library/Royal Astronomical Society, 145r Getty Images, 146 NASA, 147tl Science Photo Library, 147c ESA/LFI & HFI Consortia, 148–149 NASA/N. Benitez (JHU)/T. Broadhurst (Racah Institute of Physics/The Hebrew University)/H. Ford (JHU)/M. Clampin (STScI)/G. Hartig (STScI), G. Illingworth (UCO/Lick Observatory), the ACS Science Team and ESA, 149t Science Photo Library/Royal Observatory, Edinburgh/AATB, 149b Getty Images/Barcroft Media, 150tl, 150c, 150b & 150–151 Science Photo Library/Volker Springel/Max Planck Institute for Astrophysics, 152–153 Science Photo Library/Dr. Fred Espenak, 154 Science Photo Library/NASA/JPL/MSSS, 155 Science Photo Library/Paul Rapson

Author Acknowledgements

The elegant design for this book was by Katie Baxendale and Sooky Choi. Picture researcher Steve Behan satisfied even my ill-defined requests for obscure illustrations, and suggested many. The book's editor was Gemma Maclagan, whose concept it was, and she drove every track of the project along with infectious enthusiasm, great competence and a critical, focused eye. My thanks to them and to all of those at Carlton who helped bring the book to completion so promptly.

Paul Murdin
Cambridge, 2011